Cyril O'Brien is a retired London Fire Brigade Borough Commander, who since leaving the LFB has moved in the world of Emergency Planning, and is currently working for the London Borough of Bexley. Upon his retirement, finding additional time on his hands, Cyril made a foray into trying to write a book. His two efforts being successful, prompted him to set about this edition as his number 3. He is married to Jackie, and they have 3 very talented children, Georgina, Anne-Marie and Spencer. They have also been graced with 3 grandchildren, Coray, Anaya and Kaylan.

I would like to dedicate this book to, not only my family, Jackie, Georgina and Ahmet, Anne-Marie, Spencer and Natalie and my 3 grandchildren, Coray, Anaya, and Kaylan, but to my long-suffering companions, who have put up with my thirst for Trivia, Christmas and a little bit of Beer (and Wine), and supported me when we have travelled across Great Britain and Europe together finding Christmas Markets. Thank you, Gerard and Sue, and Dave and Jean.

Cyril O'Brien

A History of Christmas Markets through Santa's Beer Goggles

AUSTIN MACAULEY PUBLISHERS™

LONDON • CAMBRIDGE • NEW YORK • SHARJAH

A CIP catalogue record for this title is available from the British Library.

ISBN 9781398428058 (Paperback)
ISBN 9781035832910 (ePub e-book)

www.austinmacauley.com

First Published 2023
Austin Macauley Publishers Ltd®
1 Canada Square
Canary Wharf
London
E14 5AA

I would like to acknowledge one special person/entity and that is;

Father Christmas/Santa Clause/Kris Kringle/Sinter Klass/Babbo Natale/Pere Noel/Saint Nicholas, for without you none of this would have been possible.

Table of Contents

My Pick of Some Christmas Markets in the UK and Europe

These crawls will be slightly different to my first two forays into writing. I have expanded the geographical area but have maintained my enthusiasm for pubs/bars, history, trivia, ghosts and ghouls. I have also added another one of my passions Christmas.

All of these markets I have personally visited, and these are in my top ten Christmas cities. I have also included some pubs/bars that may double up as restaurants, as it is easy to forget to eat while engrossed in soaking up the atmosphere and getting in the mood for Christmas.

Those who may have an insight into my being quickly realise that Christmas is a massive part of my life. I love all things Christmas and will be sharing some of them as we go.

But, straight off the blocks, I should point out I love the Christmas story, old and new traditions, songs, films, decorations and poems.

In fact, one poem clearly sets the scene and the start of the magic, conjuring up 'visions of sugar plums' etc.

''Twas the night before Christmas. (Moore or Livingston!)'

Up until recently, this poem was attributed to Clement Clarke Moore, but in his book *Author Unknown: On the Trail of Anonymous*, Professor Don Foster gathered enough evidence to suggest it was Livingston who was the real author.

'Twas the night before Christmas, when all thro' the house,
    Not a creature was stirring, not even a mouse;
    The stockings were hung by the chimney with care,
    In hopes that St Nicholas soon would be there;
    The children were nestled all snug in their beds,
    While visions of sugar plums danc'd in their heads,
    And Mama in her 'kerchief, and I in my cap,
    Had just settled our brains for a long winter's nap—
    When out on the lawn there arose such a clatter,
    I sprang from the bed to see what was the matter.
    Away to the window I flew like a flash,
    Tore open the shutters, and threw up the sash.
    The moon on the breast of the new fallen snow,
    Gave the lustre of mid-day to objects below;
    When, what to my wondering eyes should appear,
    But a miniature sleigh, and eight tiny rein-deer,
    With a little old driver, so lively and quick,
    I knew in a moment it must be St Nick.
    More rapid than eagles his coursers they came,
    And he whistled, and shouted, and call'd them by name:
    "Now! Dasher, now! Dancer, now! Prancer, and Vixen,"
    "On! Comet, on! Cupid, on! Dunder and Blixem;"
    "To the top of the porch! to the top of the wall!"
    "Now dash away! dash away! dash away all!"
    As dry leaves before the wild hurricane fly,

When they meet with an obstacle, mount to the sky;
So up to the house-top the coursers they flew,
With the sleigh full of toys—and St Nicholas too:
And then in a twinkling, I heard on the roof
The prancing and pawing of each little hoof.
As I drew in my head, and was turning around,
Down the chimney St Nicholas came with a bound:
He was dress'd all in fur, from his head to his foot,
And his clothes were all tarnish'd with ashes and soot;
A bundle of toys was flung on his back,
And he look'd like a peddler just opening his pack:
His eyes—how they twinkled! his dimples how merry,
His cheeks were like roses, his nose like a cherry;
His droll little mouth was drawn up like a bow.
And the beard of his chin was as white as the snow;
The stump of a pipe he held tight in his teeth,
And the smoke it encircled his head like a wreath.
He had a broad face, and a little round belly
That shook when he laugh'd, like a bowl full of jelly:
He was chubby and plump, a right jolly old elf,
And I laugh'd when I saw him in spite of myself;
A wink of his eye and a twist of his head
Soon gave me to know I had nothing to dread.
He spoke not a word, but went straight to his work,
And fill'd all the stockings; then turn'd with a jerk,
And laying his finger aside of his nose
And giving a nod, up the chimney he rose.
He sprung to his sleigh, to his team gave a whistle,
And away they all flew, like the down of a thistle:
But I heard him exclaim, ere he drove out of sight—
Happy Christmas to all, and to all a good night.

Account of a *Visit from St Nicholas*, Poetry Foundation.

Before we embark on our 'crawl' around England and Europe, it may be of interest to you *The Crawler*, where all of this festive revelry began. Well, let's have a look. As you can imagine, as with the first Christmas tree, some places may dispute some versions or claims, but here is what I have discovered over my travels.

- In 1298, Vienna was given the go-ahead to hold a market (Krippenmarkt) in December, which was for all intense and purposes, the birth of the Christmas market.
- However, it was not until 1384, in Bautzen, Germany, that we had the first 'open air' Christkindlesmarkt.
- Initially, selling just meat, but over the years, progressed and expanded into selling seasonal treats and having songs and music.
- Into the 1500s, we get a bit blurry here. One school of thought is that Christmas markets got an indirect boost from Martin Luther after he pronounced that the day of Christ's birthday was probably the best day to give gifts, as opposed to other religious days.
- At the time, the churches had initially encouraged the Christmas markets to be opened and operated in the areas around the churches (hoping to encourage people into their 'services').
- But, alas, it seemed to have had the opposite effect, and the church found it was in competition with them.
- Since then, giving presents at Christmas and buying those gifts from markets flourished.

- A bit more about our friend Martin Luther, this guy had six kids.
- Devoted to his family, he thought, *how can I make this holiday season a bit brighter for the children?* (imagine there was not a lot happening for kids in the 1500s) while also emphasising Jesus' birthday and its importance.
- Martin was always struck by the beauty of the flora and fauna around the area of his home. In winter, he was especially touched by the beauty of the evergreen trees, which stood proudly upright while other plants and trees had taken their winter's 'nap'.
- In the night and dressed with freshly fallen snow, their beauty was further accentuated.
- So, one winter's day, he cut a lovely tree down and took it home.
- His idea would eventually have a dramatic effect on history.
- Explaining to his children that even during the harshest winter, this 'tree' would still flourish, was indeed a sign of hope and was similar to our faith in Jesus/God, in that even when the world is at its darkest or we are heavy with despair or grief, the tree (we) can still survive, grow and develop.
- Then, this is the good bit and explains a lot of what we do today. Martin then lit some candles and placed them on the tree. The light from the candle is a reminder of the Star the wise men followed and eventually led them to Bethlehem and that manger.

- So, in my household, I have a Star on the top of our tree (I know some use the image of the Angel that appeared to the Shepherds).
- That star remains lit until the Epiphany on 6 January. This is when the Three Kings arrived at Bethlehem, and therefore, they no longer need the bright Star to guide them; that is why all my decorations and everything come down on the 6th.
- This also ties into the 12 days of Christmas (remember that song): day one being Boxing Day (St Stephen's Day) and day 12 being the 6th of January (Three Kings Day).
- Western culture generally accepts the Three Kings being identified as Caspar (King of India: gave frankincense), Melchior (King of Persia: gave Gold) and Balthazar (King of Arabia: gave Myrrh).
- The first Christmas market in the UK was in Lincoln.
- It was born out of its twinning with Neustadt in Germany.
- So, in 1982, with just 11 stalls, our tradition (in the UK) of having similar Christmas markets began.

Hopefully, that has whetted your appetite, so let's start 'crawling' and imbibing beer, wine, Glühwein or orange juice!

So where we are going to travel to? To make things interesting, let us jump from Europe to England, then back again, and again etc.

Crawl 1:	Bath
Crawl 2:	Bruges
Crawl 3:	Rochester
Crawl 4:	Stratford upon Avon
Crawl 5:	Valkenburg
Crawl 6:	Ljubljana

Now if you have been used to reading my other books, they all take a similar format. However, as this is going to be the Christmas edition, I will stray a little, and where I need to, I will inject worldwide Christmas facts/factoids/legends etc. Also, occasionally, there may be a Christmas ghost or two.

I really hope you enjoy this book, as it's been years in the planning and researching, but do not feel sorry for me I have loved every minute of it.

So this is Christmas!

"It's not how much we give but how much love we put into giving."

– Mother Theresa

Crawl 1
Bath

"This bell is a wonderful symbol of the spirit of Christmas as am I. Just remember the true spirit of Christmas lies in your heart."

– Santa Claus, *The Polar Express*

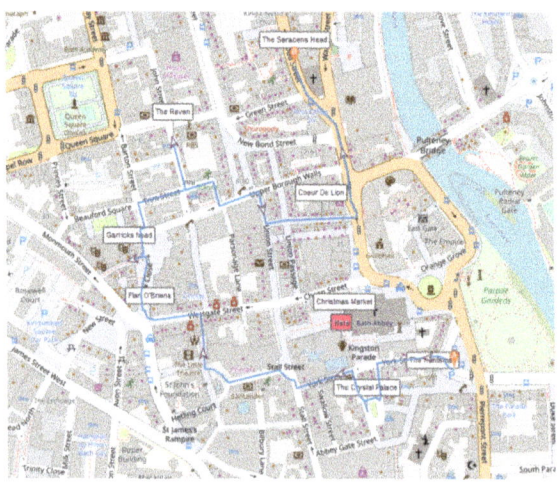

https://www.plotaroute.com/routeplanner

Christmas Market

- It is hard to keep track of new ones appearing every year, but there are over 200 Christmas markets in Britain.
- Some last for a few days, some for a few nights, while some last throughout December and into January.
- Known as the 'German Christmas markets', some follow closely to their Germanic roots while others add a distinctive English feel (Rochester with Dickens etc.).
- Bath regularly gets voted number one in Britain, even sometimes outshining the Europeans.
- Centred around the Abbey, Bath Christmas market could be the scene on a Christmas tin of biscuits, especially if it snows.
- Exhibitors keen to sell their wares come from near and far across the globe.
- Watch out and try the locally grown chillies!
- One of the great things about Christmas markets is the samples, cheese, wine, mead, beer, cake etc. You

never know what will be available to tempt your taste buds.

- Hang around or start your visit as darkness envelops the Abbey Grounds... not for the ghosts (best to keep an eye out, though!). Still, the whole scene takes on a natural wonderland feel and transports you back to Victorian times and Charles Dickens.
- As it starts to get fresher/colder, the aromas become stronger as your senses are heightened.
- Now make your way to our first pub, The Huntsman.

Pubs in this crawl: (0.819 miles)

1. The Huntsman
2. The Crystal Palace
3. Flan O'Briens
4. Garrick's Head
5. The Raven
6. Coeur De Lion
7. Saracens Head

Christmas Fact:

For over 1,200 years, garlands of Holly and Ivy have been used to adorn people's homes as it is believed to represent life everlasting. Linked to religion, the holly is supposed to represent Jesus's crucifixion crown and the berries of his blood.

The Huntsman

1 Terrace Walk, North Parade, Bath BA1 1LJ

Phone: 01225 482900

Hours:

Monday	11:00 am–11:00 pm
Tuesday	11:00 am–11:00 pm
Wednesday	11:00 am–11:00 pm
Thursday	11:00 am–11:00 pm
Friday	11:00 am–11:00 pm
Saturday	11:00 am–11:00 pm
Sunday	12:00 pm–09:00 pm

- This establishment became a pub somewhere in the late 1740s.
- It was known originally as the Terrace Wine Vaults and was owned by local wine sellers.
- In the early 1900s, the pub becomes an Eldridge Pope public house.
- It was elevated to a Grade II historic building in 1950.
- A new renovation was undertaken when the pub was taken over by Fuller, Smith and Turner Brewers.

- The Huntsman holds the title of the oldest shopfront in Bath, proud of its original keystones on the arches on the building frontage.

Route: As you leave the pub, turn left into North Parade and then take the next left York Street. Only a short crawl to the next left, which is Church Street. Follow this all the way round until you meet the junction of Abbey Green the Crystal Palace is your right.

On your way, on your right-hand side:

- The site on which the Abbey stands has had a link to religion for nearly 1,400 years.
- Back in 676 AD, there was a nunnery here.
- Nearly 100 years later, the monks moved in, and the nuns moved out.
- Between then and 1088, the building was destroyed/removed, probably by the Vikings and replaced by a Norman Cathedral.
- This was down to the efforts of the bishop of the time, John de Villula.
- Bishop John can still be seen in the Abbey on one of the stained glass windows.
- Now what could have been happening in 1137 AD?
- This year proved quite eventful with a fire damaging Bath's Cathedral, but what is really spooky (Conspiracy theorists prick up your ears) is that in 1137, not only did the Cathedral at Bath have a significant fire, but there were also significant fires in the Cathedrals at Rochester, York, Mainz and Speyer

(both in Germany), as well as the cathedrals in Dijon and Amiens! That's a lot of churches, and a lot of coincidence, although there was a terrible drought that year across Europe and England.

- In 1245, Bath was conjoined with Wells after a whole set of twists and turns, which ended up with Roger of Salisbury becoming the first Bishop of Bath and Wells after being voted in by the monks.
- This was not the baby eating bishop of Bath and Wells from Blackadder fame!
- One hundred and three years later, the Black Death hit Europe in 1348.
- It was a rapidly moving plague, and even here in Bath, no one was safe.
- In fact, the local monks lost over half their number to the Black Death.
- What also happened was a social change, as many poor people suffered and died, manual workers on farms etc., depleted in number. As shortages occurred, changes to workers' conditions could be asked for; as they had the upper hand, it almost became an employee's market.
- In fact, there is a first-hand account of the plague and its effects in the archives produced by the Priory of Rochester Cathedral.
- A new church on the existing site was started in 1499.
- However, Henry VIII had other plans, so from 1536, he disbanded churches, priories etc. (if you remember, he was a bit pissed off at the church, something about his wedding).

- This meant that around that time, the church was offered to the city of Bath for the princely sum of 500 marks.
- But the city said no thanks! And so began the decline of the building; that was until 1560.
- During this year, the then-owner of the land, Mr E. Colthurst, gave the land on which the ruins stood to Bath free of charge.

Christmas Quote:

"For me, the spirit of Christmas means being happy and giving freely. It's a tradition for all the kids in the family to help mom decorate the tree. Christmas is all about family, eating, drinking, and making merry."

– Malaika Arora Khan

The Crystal Palace

Address: 10–11 Abbey Green, Bath BA1 1NW

Hours:

Monday	11:00 am–11:00 pm
Tuesday	11:00 am–11:00 pm
Wednesday	11:00 am–11:00 pm
Thursday	11:00 am–11:00 pm
Friday	11:00 am–11:00 pm
Saturday	11:00 am–11:00 pm
Sunday	12:00 pm–10:30 pm

- In 1654, when royalty had been deposed from the throne and England was a commonwealth, this pub was known as the Three Tuns.
- Back then, there was an extra floor which provided lodgings.
- The pub we know today was named after the Crystal Palace, The (Great) Exhibition, which ran in London in 1851 from 1 May to 15 October (more about that later).

- We have a link to a naval hero here as well: Horatio Nelson (known by many titles, such as First Viscount Nelson, First Duke of Bronte, Admiral or Lord, or whatever you care to call him) who is hailed as an English hero.
- He visited Bath occasionally, and his family had ties here. Indeed, there is a plaque outside 2 Pierrepoint Street, which states, "Here dwelt Admiral Lord Nelson b. 1758 d.1805."
- On a sidebar, in one of my previous books, *A History of London Boroughs through Beer Goggles*, I recount the story of the battle of Trafalgar, and the Fighting Temeraire, a tale of daring do and well worth a read.
- The pub today still holds on to the original wood panelling, which it is said was the original timber from when Nelson stayed.
- As you can imagine, Bath has been a 'hotspot' for history since the Roman times.
- So it was hardly a surprise when in the early 1980s, Roy Wain (pub owner at the time) decided to carry out some building renovations, and he discovered something eerie and something beautiful while digging in the cellar.
- Eerily, bones of several skeletons were found.
- Beautiful; some mosaic flooring (presumably from Roman times), and it was surmised that this and the surrounding areas could be the site of a Roman villa.
- For posterity, it was protected and covered over and is still there today.

- The skeletons are still a puzzle, but remember, there is an Abbey with a graveyard nearby!
- Rather than a complete ghost alert here, there have been sightings of a cowled figure mooching about the pub, so my advice is to keep your eyes sharp and wits about you.

Route: Turn left out of the pub and continue up Abbey Street till junction with York Street. Turn left here and continue along here with the Roman Baths on your right. When you reach the T junction with Stall Street, turn right and then left into Bath Street. Follow this round to your right; this then leads into St Michaels Place. This then leads you out onto Westgate Street. Turn left here; Flann O'Brien's just down here on your right.

Roman Baths at Bath: an event, not just a swim or wash.

- The Roman Baths offered so much more than a wash and brush up.
- One could also relax while there or play games. One such game was Trigon (more about adult games later). Trigon was popular; basically, it was a three-way game of catch with multiple balls.
- The baths included wet and dry rooms, and massages were available.
- Sadly, as we know more nowadays, the baths are no longer fit for use due to the lead piping. There is, however, a hot spa that is still available.

- Not saying who is counting, but apparently, the waters contain loads of minerals, including sodium, calcium, sulphate, chloride, bicarbonate, magnesium, silicon, and iron.
- The main 'bath' is known as the Great Bath. However, it being lined with lead may not have been the healthiest option at the time. Heated water from the wells of the sacred spring fills the bath to quite a depth.

The Steamier Side of the Roman Baths

- When you look into what historians muse about the baths, it's not hard to imagine.
  - Hot and steamy… tick
  - Naked or semi-naked bodies… tick
  - Alcohol and food… tick
  - Chatting and talking politics or the subject of the day… tick More wine and food led to?
  - You guessed it! (as Donna's daughter (Sophie) said in *Mamma Mia*: dot dot dot)
  - Roman hedonism at its best
- Back in 43, Anno Domini, following the start of Claudius's attempted takeover of Britannia, Aquae Sulis (Roman Baths at Bath) was built.
- Initially set up to pay homage to Minerva.
  - Who was this goddess? Well, she was a bit of a girl!

- o Goddess of justice, wisdom, warfare strategy, and victory (makes sense as they have just set about Britannia!)
- o Minerva is also remembered for having links to the arts and trades.
- o As a mainstay of the baths, Minerva was a draw for people, seen by believers as a one-stop shop for justice. They would bring her gifts to ask for help in righting the wrong that has been done to them. This tribute could even be throwing coins in her fountain.
- o In fact, this has proved to be a treasure trove while excavating. All sorts of jewellery and coins have been unearthed, all of which started off as tributes to Minerva.

Bath has had a long association with authors Charles Dickens and Jane Austin (who has a plaque outside 4 Sydney Place, on the other side of the Pulteney Bridge) and less well-known as a Bath resident is Mary Shelley.

When Mary moved here at 19, she moved into 5 Abbey Churchyard. (She was Miss Godwin at the time, not yet married to Percy Shelley, that happened on 30 December 1816 but was short-lived as he died some six years later.)

It is believed that Mary wrote a large chunk of her book Frankenstein while living here. (Try and find the plaque!)

Flan O'Briens

Address: 21 Westgate St, Bath BA1 1EP

Hours:

Monday	12:00 pm–11:00 pm
Tuesday	12:00 pm–11:00 pm
Wednesday	12:00 pm—late
Thursday	12:00 pm—late
Friday	12:00 pm—late
Saturday	12:00 pm—late
Sunday	12:00 pm—late

- Built in around 1860ish, originally, it was known as Country Wine Vaults.
- The building was designed as a traditional Georgian property, coupled with additional aspects from the 'en vogue' Italian influences.
- The clock on the face has been given a Grade II listing.
- It was named after Brian O'Nolan.
- Who was this chap? Writing under the nom de plume of Flann O'Brien, he was a bit of a jack of all trades. He was a civil servant, writer (plays and books), and, since his death, had been recognised as a major player in the history of Irish Literature.
- He is attributed as saying about one of his motivators, James Joyce, "I declare to God if I hear that name Joyce one more time I will surely froth at the gob."

Route: Short hop now, come out of the pub and turn right up Saw Close; in about 100 m, you will see the Garrick's Head on your left next to the New Theatre Royal.

Before we get to the pub, a fire of note!
Theatre Royal Fire, 18 April 1862

- On that morning, between 9:00 and 10:00, the fire was believed to have started in the dressing room area.
- It spread really quickly and destroyed most of the interior in just over an hour.
- But the ruins smouldered on for some hours to come.
- No one was in the theatre at the time, and there were no casualties.
- The cause remains unknown (info supplied by T Morris on the Fire Service History site... many thanks).
- Straight away, designs were drawn up and work started on the rebuilding of the theatre.
- In March 1863 (less than a year later), the Theatre Royal opened its doors to A Midsummer Night's Dream.

Christmas Quote:

"For it is good to be children sometimes, and never better than at Christmas, when its mighty Founder was a child himself."

– Charles Dickens, *A Christmas Carol*

Garrick's Head

Address: 7–8, St John's Pl, Bath BA1 1ET
Phone: 01225 318368
Hours:
Monday   10:00 am–11:45 pm
Tuesday   10:00 am–11:45 pm
Wednesday   10:00 am–11:45 pm
Thursday   10:00 am–11:45 pm
Friday   10:00 am–11:45 pm
Saturday   10:00 am–11:45 pm
Sunday   10:00 am–11:45 pm

- It was built by T. Greenway in the early 1700s for the self-proclaimed King of Bath.
- Who was this King of Bath, I hear you ask, when in 1720 the King of England was George I?
- Well, Richard Beau Nash was this man, but who was he?
- Even when he was young Beau Nash was a dandy and had a penchant for fine clothes and jewellery.
- Although not very successful at his chosen profession (law), Beau found himself to be very good at gambling.
- So you can imagine, a well-dressed, confident, wealthy man (although not that handsome judging by the portraits) possessed a certain 'je ne sais quoi' to which men and women were attracted.
- From a young age, sporting a velvet coat, ruffles, diamond buckles and a diamond brooch, and soon became aware that he possessed a certain style and manner, which attracted people to him.
- In 1705, he up sticks from London and headed to Bath, which, as a city, was on the rise in 'society circles' due to the health benefits of the baths.
- His fortune took a turn for the good. After becoming friendly with Captain Webster (Master of Ceremonies), he was encouraged to take up the role of his assistant. How lucky that shortly afterwards, the unfortunate captain was killed in a sword-fighting duel, and Nash, still in his early thirties, found himself elected by the Corporation of Bath as the new Master of Ceremonies.

- Because of the recent disaster, Nash began his term by abolishing the wearing of swords and, ipso facto, the abandonment of duelling came about.

Route: Come out of pub, turn left up Sac Close and carry straight on (through the bollards) into Barton Street. At the next junction, turn right into Trim Street and continue down there till you reach the arch on your left, which leads into Queen Street. The Raven is just down here on your right.

Ghost Alert: Garricks Head Pub

As we go along, here is another bit about Beau.

It was once one of the homes of the dandy and Master of Ceremonies for Bath, Richard Beau Nash (born 18 October 1674–died 3 February 1761), and has a reputation for being haunted. The Garrick's website states that "Our two resident ghosts have yet to put in an appearance during our time at the pub. However, there are countless first-hand accounts of supernatural happenings."

The author and investigator Andrew Green (born 28 July 1927–died 21 May 2004) gave the following account of the case: the ghost here is so closely linked with the Theatre Royal next door that it is assumed to be one and the same though mysterious knocks and raps and an occasional poltergeist incident associated with the haunting affecting only the pub.

Many artists appearing in a production at the theatre have witnessed the misty grey shape in one of the boxes. Others have noticed that it travels out the window in a room just above the main bar. Frequently, the smell of jasmine accompanies these 'visitations'. The phantom is supposed to be that of a young lady who leapt to her death in the 1800s when her lover was killed in a duel with her husband. Another belief, however, is that she hanged herself in a bedroom. But why she should frequent this pub, especially the cellar, is an utter mystery. Could it be that the site of the dual was in the pub, or perhaps she may be buried beneath the barrels?

To the west of the pub, if you turn right as you come out, you will eventually reach New King Street.

Here you will find The Herschel Museum. Another famous person that is linked to Bath, William Herschel, who, while still residing in New King Street, noticed a circular type object through his telescope.

This turned out to be Uranus, and William was credited with its discovery.

- William Herschel was born in Hanover, Germany, 15 November 1738 and died when 84 years old.
- He came to Britain at 19.
- His fame stems from his discovery of the planet Uranus, but he was also the first to see Uranus's moons and named them Oberon and Titania.
- He was also a talented composer writing some 24 symphonies.
- He also became adept at making telescopes and grinding the lens. In fact, he built over 400 in his lifetime.

The Raven

Address: 7 Queen St, Bath BA1 1HE

Tel: 01225 425045

Hours:

Monday	11:00 am–11:00 pm
Tuesday	11:00 am–11:00 pm
Wednesday	11:00 am–11:00 pm
Thursday	11:00 am–11:00 pm
Friday	11:00 am–12:00 am
Saturday	11:00 am–12:00 am
Sunday	11:00 am–10:30 pm

Timeline

- Back in the 1600s, the land where the pub now stands belonged to Barton Farm.
- It was not until 1735 that this piece of land got its first placement on a map.
- Back then, the map depicted the plot as a stretch of thin garden.
- But how did we end up as a pub, or was it always a pub?
- Records show that over the years, the businesses that plied their trade out of number 7 included a grocer and a butcher.
- In the next phase, the building and streets were amalgamated as they sat on the corner of Quiet Street and Queen Street, so it was a bit blurry. The door numbers seemed to have been changed as the city grew.
- What we do know/surmise is that in 1864 Thomas Toleman (who changed the premises to a 'licensed premises') apparently lived at two addresses. Number 12 Quiet Street and Number 7 Queen Street, leading to the belief that they were indeed the same building.
- From here on in, the pub has been a 'licensed premises'.
- Various name changes have occurred since, as indeed has the ownership, names like the Wine vaults and Spirit Vaults, and even named after a brewery who took it over for it to become Fullers Wine Vaults.

Sticking with the wine links, it changed again, but this time to Hatchett's Wine Lodge.

Today, The Raven remains a traditional alehouse and is well worth a visit.

Route: Turn left out of the pub and go back down Queen Street, then turn left into Trim Street, follow Trim Street until your reach the junction with Upper Borough Walls. Turn left again here. Straight down here until we reach Union Street on your right. Go down Union Street, then take the first left into Northumberland Place; you will find the Coeur De Lion on your right a little way down.

Just before you go, north of The Raven is The Circus and The Royal Crescent. Some bits you may not know!

- You have probably seen the film *Oliver* and would recognise the Royal Crescent from one of the iconic scenes (*Who Will Buy This Wonderful Morning?*). But Bath has other crescents of note (architecturally) as well. They are Lansdown, Somerset and Camden.

- The Circus in Bath (directly north of the Raven) is as big as the area covered by Stonehenge.
- Now strangely, the Circus was built with the Royal Crescent nearby, supposedly as homage to the sun and moon.
- Thinking back to what the experts say, in that the construction of the stones had something to do with the sunrise (I can still see the crowds gathering on the morning of the summer solstice to welcome the Sun), is it conceivable that John Wood the Younger, in the mid-1760s also paid homage to Stonehenge?

Christmas Factoid:

The Legend of Sleepy Hollow: do you remember the creepy story? Well, its author, Washington Irving may have had a hand in the Christmas we see today. In his first book *A History of New York*, he introduces to the masses St Nicholas who appears to Olof Van Cortland who after seeing him decides to build New Amsterdam where he had the vision.

Ghost Alert

Every ghost story would love to have a great back story to show how the individual has been wronged, killed or killed someone or some other reason why their 'spirit' cannot pass over to the other side.

Well, one legend has it that Hatchett's Wine Lodge (one of the Ravens previous names) inherited its name after a previous licensee murdered his missus with a hatchet.

It has also been muted that the pub cellar has a bottomless well. The Lady in Grey Ghost hangs out there, and while reports of glasses being broken, eerie sounds and unexplained feelings, there is nothing definitive. But, maybe, just maybe, the ghost is not limited to one specific address and moves about. That would explain the random manifestations.

As you wend your way to our next pub, here are some bits about Ravens to whet your appetite:

The Ravens Protect the King

Legend has it that the Raven is a keystone in protecting the crown, as the portent predicts that if the six ravens leave the Tower, the monarchy will fall. So to be sure there are a few spare. At the time of writing, their names are Harris, Erin, Jubilee, Gripp, Poppy, Georgie, Edgar and Branwen.

The Raven (film 1963) starring Vincent Price, Peter Lorre, Boris Karloff and Jack Nicholson: as a kid, I remember watching *The Raven*, through the gaps in my fingers, which were covering my eyes!

Although, looking back now, it was supposedly a comedy horror film, my excuse is, well, I was young!

After the film, I decided to read Edgar Allen Poe's book and was hooked. Hard to grasp at first as a kid, but it gripped me, much like Walter de la Mare's *The Listeners*. Even the first stanza sets the mood exactly:

Once upon a midnight dreary, while I pondered, weak and weary,

Over many a quaint and curious volume of forgotten lore—

While I nodded, nearly napping, suddenly there came a tapping,

As of someone gently rapping, rapping at my chamber door.

"'Tis some visitor," I muttered, "tapping at my chamber door—

Only this and nothing more."

The Raven by Edgar Allan Poe, Poems | Academy of American Poets

This got me thinking about the Ravens at the Tower of London and how birds can be viewed differently. This was later confused even more when a Raven appeared in one of my favourite Christmas films, *It's a Wonderful Life*.

Here the Raven (real name Jimmy) seems to be the pet of Uncle Billy. Still, if you look at his screen appearances, they appear to be a subtle portent of bad news, which filmgoers may have missed.

After his first appearance, George Bailey's dad dies, followed by the collapse of the bank. The film then cuts to the lone Raven sitting in the office as George turns up to save the building and loan. Finally, Jimmy is sitting on Uncle Billy's shoulder as he misplaces $8,000.

Like Jimmy Stewart and the other actors, it was not Jimmy's first rodeo (as they say), but where could you have seen him before or after?

Before *It's a Wonderful Life*, Jimmy the Raven appeared with Stewart in the 1938 film *You Can't Take It with You*. And, strangely, he shows up again in another classic movie, *The Wizard of Oz*. Do you remember where you saw him? Jimmy is the Raven that lands on Scarecrow's shoulder.

Strange but True from the 1963 Film

- It seemed Karloff was the consummate professional, sticking to his lines etc., while others (Lorre & Nicholson) took great pleasure (much to the annoyance of Karloff, which probably made it funnier) in straying from the writer's words and ad-libbing.
- Of Jimmy, the Raven, Nicholson, who was always magnanimous about his fellow actors and crew, did not wax lyrical about Jimmy. He felt the Raven had taken a liking to crap on him! He was even supposed to have said that he hated that bird!
- If you know about actors and acting, there is always a game about trying to slip in words off script. No different here, in the film, when chanting a spell, Lorre used some mystic phrases. To the public, it sounded like an incantation in some ancient language. In fact, it was Latin and included 'Si Vis Pacem Parabellum', 'Cave Canem' and Ceterum censio Carthaginem esse delendam', not really a spell but worth googling for a laugh.
- The whole film was actually made in 15 days.
- Being in the talkies for such a long time, Boris Karloff was also in the 1935 version of The Raven (1935). However, although the two movies are loosely based on the same poem, their storylines differ.

As you go back down Trim Street, you may wonder where it got its name. Well, it was named after George Trim, who, in the 1700s, owned the land.

Unlike some other street names in and around the country. Any Trim Street links to urban slang, which would have you believe that this area was a place you may want to visit if you fancied some nocturnal horizontal or vertical exercise are not correct.

Upper Borough Walls is another historic street with lots of its buildings being listed.

As you turned left into Upper Borough walls, if you went right instead, there is a plaque resting on part of the old city walls (see above). These walls did encircle the whole city years ago, including the baths and the Abbey.

Just a short hop now to our next pub.

Coeur De Lion

Address: 17 Northumberland Pl, Bath BA1 5AR
Phone: 01225 463568
Hours:
Monday   11:00 am–11:00 pm
Tuesday   11:00 am–11:00 pm
Wednesday   11:00 am–11:00 pm
Thursday   11:00 am–11:00 pm
Friday   11:00 am–11:00 pm
Saturday   11:00 am–11:00 pm
Sunday   11:00 am–10:30 pm

This pub is famous locally for being the smallest pub in Bath.

Now, remember, you should never judge anything just on size alone.

And, while you may have to 'suffer' the small space, it is well worth it.

Previously known as Merchants Court, back in 1749, John Wood the Younger (Bath Circus) remarked it was 'entirely new'.

An oddity and a draw is the stained glass window, which depicts The Devenish Pub Company, who tried to close the pub down!

Inside is like taking a trip back in history, giving off a warm homely feel.

So what about Lionheart (Coeur De Lion)?

One of history's best-known English Kings, Richard 1, has been depicted on film, on stage, and in literature.

- Was he a good king? Well, that depends on who you speak to and what definition of a 'good king' you use.
- After being born to Henry II and Eleanor, his family life proved a little tumultuous.
- When he got older, Richard was persuaded by his mum to take the throne away from his dad!
- He eventually beat his dad in battle, and Richard ascended to the throne on 5 July 1189.
- In fact, he ruled for about a decade, and because of bravery and valour in battle, he earned the nickname 'Lion Heart'.
- In truth, he spent very little time in England, as he preferred to fight, and his battles with Saladin are the stuff of legends.
- In the end, after years of battling, he reached a peace accord with Saladin and was heading home. During a storm, his boats floundered; eventually, when trying to make it home via land (remember, at the time, he did not have many friends in Europe), he ended up

being captured and held by Leopold of Austria who done a dodgy deal with Henry VI and who in turn demanded a vast ransom (one could say a King's Ransom).

- When trying to make his way home, he tried to hide his appearance so he could pass through unnoticed. Now imagine a 6'5" red-haired monster trying to slip through without being recognised. I don't think so, and I feel the plan was flawed from the start.

- Although not crystal clear, the amount appears to have been in the vicinity of 150,000 marks.

- In today's figures, that could be worth about £17,000,000, but if you take into account inflation etc., it could slip into the billions.

- Lastly, you may have seen Disney's movie *Robin Hood*, which depicts Richard's brother 'King John' as a weak usurper not to be trusted.

- Obviously, a well-deserved character analysis; when you consider he tried to pay Richard's kidnappers to keep him imprisoned, it was quite a dysfunctional family.

Christmas Factoid:

For over 1,200 years, garlands of Holly and Ivy have been used to adorn people's homes as it is believed to represent life everlasting. Linked to religion, the holly is supposed to represent Jesus's crucifixion crown and the berries of his blood.

Route: Turn left out of the pub and continue down through the passageway till. We reach the T junction with the High Street. Here, you turn left and go straight ahead and stay on this road. Slightly bearing left as its name changes from High Street to Northgate Street and finally into Broad Street. Just down Broad Street on your right is the final watering hole on this crawl, The Saracens Head.

As we reach the High Street, just in front of you before you turn left is Bath Guildhall. Here are some bits of trivia about the place.

- The first time we came across this place in historical records was about 700 years ago.
- It was used back then as a meeting place for exiting trade guilds.
- The building was replaced circa 1625, and it was even employed to store weapons for the Civil War.
- The Civil War: Locals in Bath were split into those supporting the Royalists and those backing the Parliamentarians.
- In fact, the area became divided, with the Parliamentarians (Roundheads) having a stronghold on Bath.
- While the Royalists (Cavaliers) held Wells.
- This culminated in the Battle of Lansdown on 5 July 1643.
- The battle went back and forth, with the first strike going to the Royalists, who took a strategic bridge in Bradford.

- But the Roundheads, led by William 'the Conqueror' Walter (nicknamed the Conqueror due to his recent exploits where he successfully took and held the castles Farnham, Winchester, Chichester and Arundel in a short time), had other ideas.
- As the Cavaliers made their way to Bath, Waller moved his forces to Lansdown Hill (giving him the advantage of position).
- Seeing this, the Cavaliers had second thoughts and decided that 'fighting another day' may be the best strategy and proceeded to back away.
- However, sensing victory was at hand; the Roundheads gave chase… big mistake.
- Having lost their advantage after some back-and-forth fighting, the Cavaliers yomped up the hill into the Roundhead's guns but overcame them to secure a victory and take Lansdown Hill.
- But the victory was shallow as it cost the Cavalier army dearly in casualties.
- On your visit to the Guildhall, try and find 'The Nail'. It was here that the deals of the day happened, and it is a local belief that the term 'Pay on the Nail', may have stemmed from this very place.
- There was a significant fire here that damaged quite a bit of the building, which was back on 25 April 1972.

As we continue, you will see Pulteney Bridge on your right.

Pulteney Bridge

- What could Bath have in common with Venice, Florence, and Erfurt?
- Well, these four cities are the only cities in the world with bridges, which remain inhabited.
- Florence has Ponte Vecchio.
- Venice has the Rialto.
- And Erfurt in Germany has the Krämerbrücke Bridge.
- Back to Bath and Pulteney.
- I wonder if R. Adam, back in 1769 thought his design would not only still be in place nearly 300 years later but would be famous worldwide.
- It was built for William Pulteney, a well-off man with grand visions for Bath.
- In fact, Adam's idea for Pulteney Bridge clearly paid homage to Venice and Florence, both of which he had visited.
- I can testify (as I have made the boat trip) views of this 'wonder of the world' are best from the Weir.
- If you haven't yet been or seen The Weir, if you have seen the film Les Miserables with Russell Crowe, The Weir is where Javert jumps and meets his watery end (see picture below).

Beezer Garden's Maze

- Before we start, do you know the difference between a labyrinth and a Maze?
- Well.
- A Maze:

- This is multi-cursal, meaning having multiple paths between the start and middle.
- Generally, it is made with a degree of difficulty for the user, having dead ends etc.
- It can have more than one exit and entry point.
- Mazes can have scientific uses (think of rats, mice etc.)
- A labyrinth
  - This is unicursal, which means it has only one way in and out.
  - This is not designed to cause trepidation or be hard to complete.

So what do we have here in Bath? It is called Beezer Garden's Maze, but it is more like a labyrinth.

- So just along from Pulteney Bridge alongside the river Avon, near Pulteney Bridge, lies Beezer Gardens.
- Inside is a stone labyrinth/maze.
- It is a brainchild of Randoll Coate.
- Mr Coate, in 1984, created a bespoke 'maze', especially for Bath.
- Made out of paving stones, this labyrinth/maze must have some hidden meanings, as most other mazes he created did. We just need to find them.
- Drawn into the centre, the traveller winds forever inward until they reach a Romanesque Mosaic.

Saracens Head

Address: 42 Broad St, Bath BA1 5LP

Phone:	01225 426518
Hours:	
Monday	11:00 am–11:00 pm
Tuesday	11:00 am–11:00 pm
Wednesday	11:00 am–11:00 pm
Thursday	11:00 am–11:00 pm
Friday	11:00 am–12:00 am
Saturday	11:00 am–12:00 am
Sunday	11:00 am–10:30 pm

- This pub has been here for a long, long time and was indeed a hostelry. It was registered as a pub in 1713.
- Before then, the plot was the site of a coaching inn/hostelry.
- In 1750, when George II was on the throne, the Saracens became an overnight stop for the stagecoach known as the London Flyer.
- As ever, Charles Dickens is never far away.
- He apparently stayed here (as well as in other places) overnight in the pub's bedroom upstairs.
- Not very well known is that there is a village not far from Bath called Pickwick.
- Now, if you know about Dickens' writing escapades, you will remember his first book, *The Pickwick Papers*.
- This old pub is even talked about in that book by name.
- Curiouser and curiouser, the main character Mr Pickwick is believed to have been modelled on a local man, Moses Pickwick.

Ghost Alert

As you can imagine (again!), in such an old establishment, the presence of a ghost or two is a given.

So who do we have? Well, there is a 50/50 chance of which one you will see.

1. In the early 1800s, a parishioner at St Michael's fell to the ground outside of the pub; no one is quite sure if he was going in for a pint or two, as he died before

they found out. He (the grey ghost) seems to favour the bar as his preferred haunting site.

2. Another apparition is that of Scottish Al, but apart from a few sightings, however, there is little known about him or his demise. Makes you wonder how they knew he was Scottish!

Christmas Factoid:

Ebenezer Scrooge's famous line 'Bah Humbug' almost never existed. Charles Dickens' initial choice was 'Bah Christmas'. Glad he changed it.

So, as we leave Bath, pat yourselves on your back. Well done, another one completed, but to finish off here is some additional trivia.

Bath: Did you know?

- As late as 2007, a load of coins from Roman times was unearthed in the bowels of Gainsborough Spa (Beau Street). Hidden in several bags was a large haul of coins dating back from 30 BC. The find is now on display at the Roman Baths.
- Hitler wanted to demoralise the British population during WW2. In 1942, the Luftwaffe hatched a plan to hit and destroy the British Morale. His cunning plan was to use an old German tourist guide, the

Baedeker, written in the 1800s and highlighted 'must see' places in Britain.

- His hopes were that by destroying these icons, he would break the British Spirit and Heart.
- Was it successful? No! Although there was some damage, the Baths (Roman and Georgian) remained mostly intact.
- What is special about 2 May 1840? Well, here in Bath, something quite historic took place.
  - A letter was sent from the Bath Post Office, with a Penny Black attached!
  - Not earth-shattering news until you realise this was four days before the official launch of the stamp.
  - I wish I had one of these, the Penny Black. In pristine condition, you may get hundreds of thousands of pounds for one and to think it all started here in Bath.
- In the early 1900s, Eagle House was given over as a safe haven for the ladies who protested who had been imprisoned for protesting for womens rights, following their release, namely E & C Pankhurst, A. Kenney.

Just Saying:

On Christmas Eve in 2001, the Bethlehem Hotel had 208 of its 210 rooms free.

Crawl 2
Bruges

"There's a certain magic that comes with the very first snow.
For when the first snow is also a Christmas snow, well,
something wonderful is
bound to happen."

— Frosty the Snowman

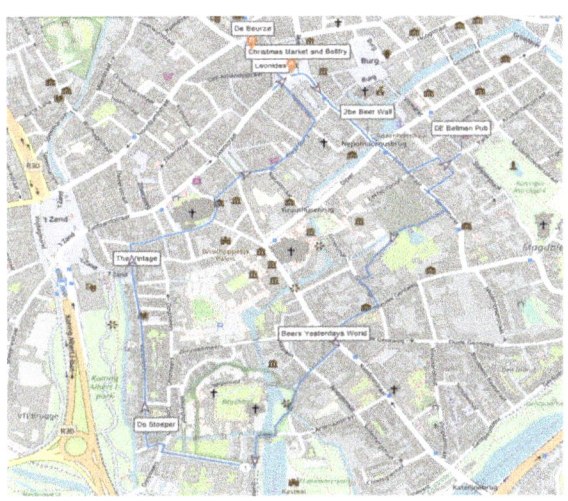

https://www.plotaroute.com/map/1888508

Christmas Market

- This was the first Christmas market I visited, and I have been back many times. It is my 'old favourite'.
- As you arrive in this city (and hopefully with the snow), you are walking into a quintessentially picturesque image of Christmas.
- The scene would not be out of place on a special Christmas card.
- Surrounded by tall old buildings, a market square with an Ice Rink, cobbled streets, a Gothic tower, horse-drawn carriages inviting you to explore, and aromas of Christmas everywhere… it is magical.
- The whole city engages in transforming Bruges into a city of Christmas. Shops and bars are all decked in their finest Christmas attire.
- If you need to get into the spirit, you would be hard-pressed to not be 'Christmatised' by the time you leave here.
- The last time I went, we were lucky to also have an ice sculpture exhibition. If it's on when you get there, make sure you leave time to visit it.

- Bruges has the lot. From traditions like St Nick, who arrives by boat from Spain to special meals and foodstuffs, including their version of the Yule Log? To scenic views with amazing chocolates and beers.
- We will also pass another Christmas market on our crawl, but more about that later.

Christmas Quote:

"We consider Christmas as the encounter, the great encounter, the historical encounter, the decisive encounter, between God and mankind. He who has faith knows this truly; let him rejoice."

– Pope Paul VI

So where will be going? On this crawl around Bruges, we will be stopping off at:

1. The Beer Wall (Wollestraat 53, 8000 Brugge, Belgium)
2. The Bellman Pub (Jozef Suvéestraat 22, 8000 Brugge, Belgium)
3. Beers Yesterday's World (Wijngaardstraat 6, 8000 Brugge, Belgium)
4. Do Stoeper (Oostmeers 124, 8000 Brugge, Belgium)
5. **Another Christmas market in Zand Square
6. The Vintage (Westmeers 13, 8000 Brugge, Belgium)
7. **Another Christmas market in Simon Stevinplein
8. Leonidas Chocolate Shop (Confiserie Dalipan Steenstraat 4 8000 Bruges)
9. De Beurze (Markt 22, 8000 Brugge, Belgium)

This crawl is just under two miles.

So, before our first drink (unless you had some Gluhwein at the market) and as we are standing in front of the Belfry, just some bullet points before we head off.

- It was built in thirteenth century.
- The 83 m high Belfort (Belfry and Carillon) gives fantastic views from the top.
- This tower is one of the three highlights on the Bruges Skyline.
- The other two are the towers of the Church of Our Lady and Saint Saviours Cathedral.
- If you make it to the top (all 366 stairs), on your way, you may come across the carillon bells keyboard. (A carillon is a ring of bells played via a keyboard or by an auto-mechanism like a piano roll.)
- The carillon in the belfry tower holds at least 47 bells.
- Bruges has its own Carillonneur, who plays the bells at specific times throughout the week.

Route: As you face the Bell tower turn left along market, then follow on round to the right and continue down Wollestraat until just before the bridge, and you will see our first bar on your left The Beer Wall.

As we make our way down into Wollestraat, had you taken the left pedestrian street called Briedelstraat, rather than carry on, it would have led you into Burg Square, and tucked away in the corner is the entrance to the Basilica of the Holy Blood.

So, while we are making our way to The Beer Wall, here are some titbits about the Basilica.

You may be asking what actually makes a church a Basilica.

The word itself comes from Greek, translating to 'royal house'. However, within the Roman Catholic religion, it is a term to describe a church, which has been recognised and given additional privileges by the Bishop of Rome (Pope). For this church, it happened in 1923, while Pius XI was in charge at the Vatican.

So what do we know about the Basilica of the Holy Blood?

It is divided into two parts (Lower and Upper).

The Lower Basilica of Saint Basilius

- This is the only Romanesque-style basilica/church in Western Flanders.
- It was built the 1100s and was named after Basil the Great (who was made a Saint soon after his death on 1 January, AD 379, and was bishop of Caesarea (Asia-Minor)).
- Basilica houses numerous relics, even one purporting to be from Basil himself.
- Some impressive statues are here to admire, and some are paraded during the Holy Blood procession.

The Upper Basilica

- It was built in the Gothic style at the end of the 1400s.

- Some of the great stained glass windows are nearly 200 years old.
- The Gothic upper basilica was built at the end of the fifteenth century. Noteworthy here is the painting and fresco on the altar wall.
- It not only shows the death of Jesus on the cross but the lower fresco recounts the journey of his Blood to Bruges.

So how did Jesus's blood end up here in the church in Bruges?

Now every good Catholic knows that after his death on the cross, where he had been speared and made to wear a crown of thorns, Joseph (of Arimathea) washed him and wrapped him in a shroud before taking him to the tomb (from where he rose again). However, the blood-soaked rags were not disposed of but rather kept for posterity by Joseph in Jerusalem.

From here, we have two versions of history/legend.

- Thierry, a Flanders Count in the twelfth century brought it to Bruges, but this version has very little substance to it.
- Another version of how it reached Bruges (favoured by many) is that following the demise of Constantinople in the thirteenth century, many of the treasures and relics were, how shall I say, relocated! And some to this side of Europe.
- How did it get to Constantinople in the first place from Jerusalem I hear you ask?

- Well, there was a lady, Saint Helena, who was Emperor Constantine's mother and she is believed to have brought lots of treasures etc. from from Jerusalem to Constantinople. See, there is a plausible link!

Okay, back to the hunt for beer.

Christmas Quote:

"And when the Lord Jesus has become your peace, remember, there is another thing: good will towards men. Do not try to keep Christmas without good will towards men."

– Charles Spurgeon

The Beer Wall 2BE

Wollestraat 53, 8000 Brugge, Belgium
Hours:

Saturday	—	10:30 am–8:00 pm
Sunday	—	10:30 am–7:00 pm
Monday	—	10:30 am–7:00 pm
Tuesday	—	10:30 am–7:00 pm
Wednesday	—	Closed
Thursday	—	10:30 am–7:00 pm
Friday	—	10:30 am–8:00 pm

This place is impressive. As you enter through the archway, you are greeted with a wall of beer. The choices are astounding, and what was awesome was that every beer has its own special glass, which is a nightmare for someone like me who collects nice glasses for my home bar!

- The site sits on the old Mayors' house, which was here in the 1400s.
- Back then, this was where the Holy Blood (which we spoke about earlier), was kept.
- Nowadays, there is a shop selling a wide array of beers, crackers and chocolate.
- What is truly amazing is that there are over 2.000 beers to choose from to either take away and drink at your leisure or get a table and drink here, overlooking the bridge and canal.

Route: Turn left out of the pub and cross over Nepomucenusbrug. Once across, go straight down Eekhoutstraat and take the first left into Geerolfstraat. Go to the T junction, turn right then first left into Gevangenisstraat, and De Bellman pub is down here on the left.

As you go over the bridge, you will see a statue of St John of Nepomuk.

Now here are some interesting facts about St John. Saint John is also a key figure in Prague, Ljubljana and other parts of Europe and is important to me, as an ex-firefighter. Where to start. Well, the statue itself depicts St John, but that does not tell the story of his life and death. For that, you need a bit more information.

- John of Nepomuk was born in the mid-1300s and died aged 48.
- He was made a Catholic Saint in the Czech Republic (then called Bohemia) soon after his death. Why?
- He was confessor to King Wenceslaus IV (not the Good King Wenceslaus, that went out on the feast of Stephen!), a position he also held for the king's second wife, Queen Sofia.
- At the time, the king felt his wife was being unfaithful to him. He was sure she had confessed her indiscretion in the confessional.
- So the king summoned John and asked him to reveal his wife's secret confession.
- John refused! Well, not being happy with this answer, the king threatened that unless he told him, he would have him tortured.
- John stuck to his guns (good man that he was) and told the king that the seal of the confessional was sacrosanct and he could not talk of any matters discussed within the confessional.
- True to his word, King Wenceslas IV tortured him with fire, but as John held out, the king began to panic about how this would look (torturing a priest) and released him.
- But now, worried that John would tell everyone about his torture at the hands of the king, he gagged him, apparently forcing his body into a goatskin, and then threw it into the Vltava River, where he drowned.
- St John of Nepomuk is believed to be the first martyr to have died to protect the seal of the confessional.

- Prague is well worth a visit, as is St John's statue. There is a local legend that has visitors touching the statue for good luck. But, if you do go, be wary of the image of the dog!
- So why is he here in Bruges? Well, what shocked me is that he is also in Ljubljana as well.
- In Ljubljana, the statue depicts the event of him being thrown from the bridge.
- This is outside the St Florian's Church (patron saint of firefighters). Now do you see the link?
- Because of the manner of his death, St John seems to have inherited the mantle of protector of bridges, protector against floods, and believe it or not, drowning!
- So don't be surprised if you see him protecting a bridge on your travels across Europe.
- One distinctive feature on many of these statues is that above his head are five stars.
- Legend has it (and Prague does like a legend) that on the night he was tossed into the Vltava, these five celestial bodies clearly stood out in the night sky.
- It is believed these stars led some locals to locate where John had washed up.
- The stars may also be paying homage to the five wounds of Jesus Christ sustained during his crucifixion.
- Some statues also have him holding one finger to his lips to symbolise the silence he kept.

Right, you are probably near our next watering hole.

The Bellman Pub

Address: Jozef Suvéestraat 22, 8000 Brugge, Belgium
Hours:

Monday	Closed
Tuesday	Closed
Wednesday–Saturday	11:00 am–1:00 am
Sunday	Closed

This is a comfortable pub, which is perfect to drop into and have one on the way. I love the collection of pipes and beer tankards, which adorned the place, making it quite quirky.

Route: Come out of the pub and turn right and head back down the way you came until you reach Eekhoutstraat, then turn left into Garenmarkt, and follow this road until the junction with Niewe Guentweg. Here, we go right and stay on this road until we reach the junction of Katelijnestraat. Left again and then right into Wijngaardstraat and the next pub is down here on the left.

As you wander down the route, while there is nothing specific to note, apart from a museum, here are some facts about Bruges to help you on your way.

A Potted History

- The area around the Bruges region has been populated since the Romans expanded across Europe.
- But the first appearance of Bruges in the written word was in the 800s. It was likely to have been a derivative of the Germanic term 'brugi', which loosely translates as a 'mooring' place.
- Although not directly on the coast, Bruges has always had a close link to the sea, and its canals are fed from the sea.
- In olden times, there was a marine channel that fed Bruges, but when they got silted up, the Brugge-Zeebrugge Canal (built between 1896 and 1907) re-established the marine links.
- Indeed these watery links played very well for Bruges in its middle-age development and its status as a port.
- The town itself became fortified and turned into a mighty political stronghold. This was mainly because the Flemish Counts resided there and controlled Flanders.
- As a port, in the 1200s, this city turned into a leading trade centre in the North West corner of Europe.
- Bruges quickly became a desirable place to live for merchants all over Europe. This was further

fermented when the world's first recognisable stock exchange started here in Bruges.

- Now some disagreement lies here. (Controversy, there's a surprise!)
- What we do know is both Bruges and Antwerp were the hubs of trade (although Antwerp eventually had the first purpose built stock exchange in 1531), but with the Venetians and Germans descending on the city in the 1300s, they needed a place to stay and have a bit of fun and guess where this happened... pubs or inns as they were called back then.
- This led to the inns being the first trading exchanges, simply because while people met and drank, they struck deals; the rest, as they say, is history.
- One of these pubs was the Ter Buerse Inn in Bruges. Since the 1200s, it has been run by the family Van de Beurse.
- This inn then became the place to be and to trade. Indeed, it is believed that the Van de Beurse name is the original basis for the French word Bourse, which translates as Stock market (NB Beurs in Dutch means exchange).
- Something to look out for is that when you reach the turning into Katelijnestraat, we turn left to the next pub. However, if you detour and turn right a short way up, on your right is the Church of Our Lady Bruges.
- This is a church full of artworks including, the 'Madonna and Child' created by Michelangelo (a very special piece with a great historical story).

- The church's tower (spire) is part of the trio of towers which help form the skyline of Bruges.
- The other two are The Belfry (where we started) and St Saviour's Cathedral, which we will discuss later on this crawl.
- Stretching 115.5 metres into the sky, this wonder holds the title of the second tallest brickwork tower in the world. The tallest is St Martin Church in Landshut, Germany, which stands a whopping 130.6 m.
- Michelangelo's 'Madonna and Child' (made from a single slab of marble!) was commissioned by Mouscron (a wool merchant from Bruges, but who had a trading base in Italy) decided to send it back to Bruges.
- The sculpture is a mere 4 feet 2½ inches high, but it is stunning.
- As with everything, if there is a back story, it makes it more interesting, which is no different for the 'Madonna and Child' (Madonna of Bruges).
- This is the only known work by Michelangelo to have left Italy back then, and it was stolen not only once but twice.
- Mouscron kept the statue for himself and placed it in his own chapel. Here it remained until 1794 (the French Revolution) when it was nicked by the French.
- It remained in Paris for 22 years before being returned to Bruges.

- Another period of stability ensued, that was until the Nazis turned up in 1944 and took it for placement in Hitler's museum.
- Enter the Monuments Men (George Clooney, Matt Damon, John Goodman, Bill Murray, Hugh Bonneville). Although this is a film and these are the actors, the true story is that the characters they played, rescued tons of artwork and returned them to their owners, including the Madonna of Bruges.

Back to our next pub.

Christmas Quote:

"Once in our world, a Stable had something in it that was bigger than our whole world."

– C.S. Lewis

Beers Yesterday World

Address: Wijngaardstraat 6, 8000 Brugge, Belgium

Hours:

Monday	—	4:00 pm–1:00 am
Tuesday	—	4:00 pm–1:00 am
Wednesday	—	4:00 pm–1:00 am
Thursday	—	Closed
Friday	—	4:00 pm–1:00 am
Saturday	—	2:00 pm–1:00 am
Sunday	—	4:00 pm–1:00 am

Looking for a place with character, well this has it in spades. Here is a bar in an antique shop.

Good Beer Guide Belgium 8 describes this place as 'a sort of junk shop with hats'. The owners go with 'vintage shop and café pub', but whatever you call it, this place has personality—oh, and around 50 beers on a decent list.

So grab your beer, sit back and take in your surroundings before we go on our way.

Route: Turn right out of the pub and go straight, continue on the pedestrian area and follow Wijngaardplein over the bridge into Begijnhof and turn right into Begijnenvest. Follow this round to the right and continue to the junction at the end. Here turn right, then first left and the pub is on your left. (Phew!)

Lake of Love

- As you pass over the bridge, the expanse of water you see, which is actually a reservoir known locally as Lovers Lake.
- Situated inside Minnewater Park, which is roughly translated from the Dutch language meaning 'love water'.
- The story (folklore) tells of the 'doomed' love story of young 'Minna'.
- Now Minna was in the unenviable position of being set up by her father for an arranged marriage with another.
- But Minna was in love already.

- As with many 'love stories', the person you love may not be the person you are expected to be with.
- Minna loved a soldier/warrior who was not far away but belonged to another tribe! (See a pattern forming here!)
- So to avoid the arranged marriage, Minna ran away through the forest to meet up with her lover.
- However, in her desperation, she ran so hard and for so long, and by the time she did find her 'beau', she died in his arms from exhaustion.
- Now the legend has it that if you stroll across the bridge (remember there is one at each end of the 'lake') with your lover, if you Embrace and Kiss, you will be blessed with a love which lasts forever. Ahhhh!
- Also, it is a great place to see wildlife, especially swans, which, as you know (or not), are Bruges city emblem.

NB. You may be lucky and also find an artificial Ice Rink put up in Minnewater Park for the Christmas Period, but you need to check the Bruges Website for the most up-to-date information.

Some factoids: What is Bruges famous for?
Well, for my part, there are four things:

1. Chocolate (you have to try this)
2. Lace making
3. Diamonds

4. And being labelled the world's most Boring City (except when its Christmas markets, then it's one of the best)

- A little known fact is that French fries or American Fries were actually invented in Belgium.
- The story goes that the fishing villages along the Meuse River (in Belgium), used to fry their catch. However, in the deep of winter, when the River froze over, there was no fish to catch. The people used to fry potatoes instead.
- This was discovered by the Americans in World War I, coupled with the fact that the predominant tongue in the area was French, as they say, the rest is history.
- Bruges, aka 'Venice of the North' (due to all of its canals), would be a nightmare to navigate and get about. How did they overcome this? Well, for a little city, there are over 80 bridges!

Some Christmas Bits

- As Belgium belongs to the group of low countries, some of the traditions of these areas naturally travel over time across borders.
- Now you may be aware of the term 'fill your boots', but this gives a whole new spin on the saying.

- Who we know as Father Christmas. Here in the Netherlands, he is known as either Sinterklaas or Saint Nicholas.
- St Nicholas's Saint Day is celebrated on 6 December. So here on St Nicholas Eve (5 December), families and friends gather to party. This is known as Pakjesavond, and gifts are also exchanged.
- It is the belief here that Sinterklaas arrives from Spain on his flying white horse with his helper Zwarte Piet (Black Peter) and will leave gifts for the good children, but the naughty children have another fate, they get taken away back to Spain!
- In preparation for this event, kids put their shoes in front of the fireplace in the hope that they will be filled with gifts (not unlike our Christmas Stockings) and treats.
- Another twist on this tale is before going to bed, Children will place their (or a spare) shoe down the chimney, already filled with carrots etc. (as a gift for the night visitors). And then sing songs. The next morning they wake up very early to find a gift, or biscuit inside.
- As for Christmas, it all begins with the arrival of Saint Nicholas in November along with his companion, Black Peter. They arrive from Spain in a boat docking in one of the Dutch ports.
- Apparently, these pair live in Spain for the rest of the year.

Is There a Change Coming?

- As you can imagine, although the legend has it that Black Peter is a Moor from Spain and appears to have dark skin because he has been inside Chimneys for a long time. Nowadays, some are concerned about stereotyping and although the current thinking is not to change the tradition, how long this will last, who knows?

Christmas Quote:

"The very purpose of Christ's coming into the world was that he might offer up his life as a sacrifice for the sins of men. He came to die. This is the heart of Christmas."

– Rev. Billy Graham

De Stoepa

Address: Oostmeers 124, 8000 Brugge
Hours:

Monday	—	Closed
Tuesday	—	11.45 am–2:00 am
Wednesday	—	11.45 am–2:00 am
Thursday	—	11.45 am–2:00 am
Friday	—	11.45 am–2:00 am
Saturday	—	11.45 am–2:00 am
Sunday	—	11.45 am–2:00 am

The Stupa is a characterful restaurant and a great little conversation bar just on Bruges city edge. Lots to eat and plenty of beers to choose from.

But what is a stupa?

The stupa is an ancient word, which comes from Sanskrit meaning heap. Generally associated with Buddhism and used to describe a type of building, it is normally associated with being a burial place or where religious artefacts or relics are placed.

Route: Turn left out of the pub and take the next right into Westmeers and follow it all the way up until you see The Vintage on your right, on the corner Saint Jan in De Meers.

Not much to see down this slip of a road, so some more titbits about Bruges and Christmas.

- Not to be outdone (however, Iceland, I think they have 13), in Belgium, there are two Father Christmases'.
- For the Dutch, there is Sinterklaas (St Nicholas).
- For the French, there's Pere Noel (Santa Claus).
- Also, Belgians celebrate Advent. They make wreaths from fir, in the shape of a cross (X), with a candle at each end of the arms. One candle is lit each Sunday in Advent.
- Like the English, they decorate using trees, lights, wreaths etc.
- On Christmas Eve, the family and friends gather for a special meal, after which presents are put under the tree for the following day.

- On New Year's Eve, guess what? A large meal is prepared and shared with family and friends, and more gifts are given.

- On the 6th of January (The Epiphany), everyone celebrates the Adoration of the Magi (The Three Kings or The Three Wise Men).

- Here the kids get done up like the Three Kings and go door to door, similar to carol singing or trick or treating.

- There is also the delicacy known as the Three Wise Men Pie, known locally as the 'Galette des Rois'. This cake has a crown on its top. Tucked away inside is a hidden treasure. Whichever child finds the treasure is adorned with the crown and is king or queen for the day.

Christmas Quote:

"I once bought my kids a set of batteries for Christmas with a note on it saying, 'Toys not included.'"

– Bernard Manning

Off we go again.

The Vintage

Address: Westmeers 13, 8000 Brugge, Belgium
Hours:

Monday	—	Closed
Tuesday	—	Closed
Wednesday	—	2:30 pm–12:00 am
Thursday	—	2:30 pm–12:00 am
Friday	—	2:30 pm–2:00 am
Saturday	—	2:30 pm–2:00 am
Sunday	—	2:30 pm–10:00 pm

This pub is located near the centre of Bruges, close to the railway station and not far from the Christmas markets.

Another characterful cosy pub, helped by the Vespa bike, which is hanging from the ceiling!

Lots of choices of beers, even ones that the Trappist monks are responsible for. But, beware, some of these beers are extremely strong. Coupled with the cold outside and the warmth inside; you will not be shocked to know that some food may be essential to help maintain an upright position.

Route: Come out of the pub and continue down Westmeers until you reach the junction with Korte Vuldersstraat, where you turn right. Turn next left into Lendestraat, then right into Zuidzandstraat and continue straight into Steenstraat. Straight down here until we near the market square where we started, and you will see Leonidis on your left, a great chocolate shop. Once you have had your fill here, carry on into the Markt square, where you bear left, and

De Beurze is in front of you on the opposite side of the Bellfry.

Another Christmas Market!

Yet again, the Christmas spirit hits Bruges. Since 2021, there is going to be another Christmas market at Zand Square. So, if you want to visit this as you get to the end of Westmeers, instead of turning right, turn left into t'Zand, and follow to Zand Square.

This one is set up for families and children and will, in all probability, grow year after year.

So, if we get back to our crawl when you turn right off Westmeers and follow the route, you will end up in Steenstraat, then on your right-hand side is a lovely Cathedral.

Christmas Quote:

"The main reason why Santa is so jolly is that he knows where all the bad girls live."

– George Carlin

St Saviours Cathedral

- This cathedral is the main one for Bruges.
- Staring life in the 900s as a Parish Church and not rising to the position of a cathedral until the nineteenth century.
- Following the appointment of a new Bishop in 1834 (after Belgium's independence in 1830), St Saviours became a cathedral.

- Its visage has changed since then, helped in no small part by a fire in 1839, which caused the roofs to collapse.
- During the renovation, a new tower was built following the designs of Robert Chantrell (a famous English architect) and the building of its new tower.
- What they managed to do was to incorporate part of the twelfth century building into the base of the tower, thereby having some continuity.
- Unhappy with Chantrell's final building, the authorities added a 'topping' to the tower (they thought it was too plain and flat), so that gave it a peak!
- The top of this peak now stands an impressive 324 feet high.

Another Christmas Market

A few minutes away from the main market where we started, there is another Christmas market, albeit a little smaller. It is in Simon Stevin Square (Simon Stevinplein). When the Christmas market stalls are here, the area and trees are decorated with lights making it very picturesque.

We then carry on down Steenstraat until we reach Leonidas.

Now there are plenty of chocolate shops in Bruges, but this was the first one I tried over 10 years ago, and I always go back. I simply love the chocolate Pringles!

Now after loading up with chocolate, we go into the square, and opposite the Belfry is De Beurze.

De Beurze

Address: Market 22, 8000 Brugge, Belgium
Hours:

Monday	—	9:00 am–10:00 pm
Tuesday	—	9:00 am–10:00 pm
Wednesday	—	9:00 am–10:00 pm
Thursday	—	9:00 am–10:00 pm
Friday	—	9:00 am–10:00 pm
Saturday	—	9:00 am–10:00 pm
Sunday	—	9:00 am–10:00 pm

De Beurze is a nice spot if you fancy some Belgian food. Choices include Stews, Steaks, and of course, beers. The staff were friendly and attentive when I went. However, if you're inclined to research, customer reviews are mixed.

During my visit, I was told that there were a lot of fake reviews for a lot of places on the square, and the guess was, that it was from rival restaurants. What I suggest is to take a look, see how you are greeted and make your mind up on the night. There are plenty of other places in the square.

What was lovely was the real fire, which made the cold night seem very far away.

Okay, as we come to the end of our crawl, let's have some more Christmas bits.

- Now at Christmas, the airwaves seem full of hallmark Christmas films. But did you know they only take two weeks each to film?

- Now, have you ever seen a 'floating' Christmas tree? Well, in Brazil, they have! In fact, they set a world record for it. Standing at an impressive 85 metres, give or take a whisker. (That's equivalent to a 28-storey building), add to that millions of lights, I bet it was really impressive when the record was set in 2007.

- Pope Julius I, in the middle of the fourth century, decreed that Christmas would be celebrated on the 25th of December.

- The kick is, is that it was not that the 25th of December was Christ's birthday, but it was chosen to compete with Saturnalia, which was a pagan celebration.

- In fact, the very first record of Christmas was in 336 AD.

- Did you know that the well-known Christmas song Jingle Bells, was actually the first song to be played in outer space? This all happened in 1965 aboard the Gemini 6A. However, I cannot verify if any alien race beat that record.

- Have you ever considered where the idea of putting stockings above the chimney breast on Christmas Eve came from? In America, they hold to the legend that if you hang up stockings on that night, Santa will come and fill them with gifts.

- Now, that legend goes back to where it is believed that a man who had three daughters was distraught. Because he was so poor, he would be unable to afford their dowries, so they would remain single.
- But learning of his plight, St Nick climbed down the chimney on Christmas Eve and filled the daughter's socks with Gold coins, providing them all with a dowry.
- When they awoke the next day, all were very happy, and the girls all managed to find a suitor.
- Most people have seen the NORAD tracker, which has traced Santa's journey on Christmas Eve since 1958, but did you know it was through pure chance and the actions of one man which allegedly gave rise to this Christmas magic? Apparently, in 1955, a child thinking they were calling Santa (their town's local paper was highlighting a Santa-style event), decided to call and speak to Santa. But they misdialled and got through to Command at Air Defence.
- Here is where an unsung hero comes to the fore. Let me introduce you to Colonel Harry Shoup. It was his quick thinking and recognition of the need to maintain this child's 'belief' he told the child that his team at Air Defence would guarantee that Santa would be safe on his journeys that night. Thus, magic was created.

- Probably not too big a surprise, but if you followed the song The Twelve Days of Christmas to the letter, you would actually give a total of 364 presents.
- The Canadian Authorities have given Santa an actual postcode. Although it lists the North Pole in the address, the coding that follows HOH OHO Canada ensures it gets delivered and letters are all answered.
- Back in 2019, the Spanish Hotel Bahia decorated what is reputed to be the most expensive Christmas tree in the world, with ornaments and baubles from Bvlgari Cartier Louis Viton etc., the final cost came in at nearly £12 million.

Just north of where we are, there is a place you make like to visit, if you like ghosts.

Spanjaardstraat 17, 8000 Brugge Ghost Alert

- In Bruges, there is a notorious house that is known locally as Het Spookhuis, its real name being Den Noodt Gods.
- As with most ghost stories, we have to delve back into history to see where the catalyst for the haunting started.
- So back very early nineteenth century, next to the River Reie stood a nunnery. Do you know the difference between a nunnery, convent, monastery etc.?

- Well, generally, a monastery or nunnery is an enclosed community of people living by the regimes of the religious vows they have taken.
- On the other hand, a convent or friary is a group of believers who have given up any earthly possessions and live very simply.
- So now, back to the story, on the other side of the Reie was an Augustinian monastery.
- This is believed to have had a tunnel which joined the two, but if we are to believe the stories, the tunnel was never used, and the Nuns and monks used to traverse the River using the bridge.
- As we know, most religious orders especially Nuns, Monks and Priests take vows of chastity. No sex!
- Legend has it that there was a young (soon-to-be) Nun (Novice) named Hortense.
- As is their way, although they live a life basically without sin, feel the need to 'confess' regularly.
- So there was a young Priest who would hear the confessions of the Nuns. Can you guess where this is going?
- Sure enough, his hormones got the better of him, and he fell for Hortense.
- No matter how much he threw himself into prayer and worship, something kept popping up! (ohh! err!)
- So after discovering the secret tunnel, the Priest sneaked off to find Hortense and tell her of his feelings.
- When he found her, it became apparent that they were mutually attracted, and he made an advance to kiss

her, but she held firm and said no. She asked him to leave but return the next night, which would give her time to think.

- Night after night, he returned, his ardour growing and as was his frustration.
- Hortense was still confused.
- At the end of his tether, becoming ever more frustrated, he finally snapped, killed Hortense, and smuggled her body out of the nunnery through the tunnel.
- Not long after her disappearance, the nuns started seeing apparitions floating and screaming through the nunnery halls.
- This caused some of the nuns to leave the nunnery.
- It was even reported that the pair had been seen in the building!
- Fast forward 200 years or so, and a family from England was now resident on the property, and the nunnery was now closed.
- We know the family were called the Unlackes (not unlucky, but might as well have been!).
- They were subject to countless sightings, so much so that they called in a spiritualist to communicate with the spirits.
- But this did not go to plan!
- Mr Eglinton had such a bad experience and nearly went into a coma (not sure how you can nearly go into a coma, but there you are!).
- But while he was suffering, he did manage to chat with the Nun and the Priest (there's a surprise!), who

both confirmed that it was the priest who killed her in a fit of rage and buried her body.

- So if you see the Nun or frustrated Priest floating by, you know why.

Christmas Quote:

"It's Christmas Eve. It's the one night of the year when we all act a little nicer, we smile a little easier, and we cheer a little more. For a couple of hours out of the whole year, we are the people that we always hoped we would be."

– Frank Cross
Scrooged

So back to England now for another Christmas market!

Crawl 3
Rochester

"From the towns, all inns have been driven; from the villages most. Change your hearts, or you will lose your inns, and you will deserve to have lost them. But when you have lost your inns, drown your empty selves—for you will have lost the last of England."

– The Hilaire Belloc blog.

Rochester

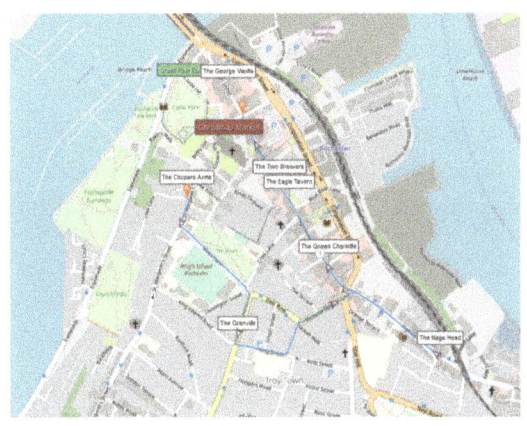

https://www.plotaroute.com/route/1888592

Christmas Market

(Castle: Castle Hill, Rochester, Kent, ME1 1SW)

Christmas markets: These will take place over the weekends of November and December.

The Dickensian festival takes place on the High Street between the bottom of Star Hill and the Medway Road bridge: there additional events happen in and around the Castle and the Cathedral mainly centred on Medway Visitor Centre, postcode ME1 1LX.

The Dickens Christmas Festival will take place over the weekend of December 4–5 with a Victorian theme taking over Rochester High Street between 10:00 am and 5:30 pm on both days.

Pubs on this crawl!

The Coopers Arms	10 St Margarets Street Rochester ME1 1TL
The Granville	83 Maidstone Rd, Rochester ME1 1RL
The Nags Head	292 High Street, Rochester ME1 1HS
The Queen Charlotte	159 High St, Rochester ME1 1EH
The Eagle Tavern	124 High St, Rochester ME1 1JT
The Two Brewers	113 High St, Rochester ME1 1JS
The George Vaults	35 High St, Rochester ME1 1LN

Crawl is About 1½ miles

Rochester

It used to be a city! Believe it or not, from 1211 right up to 1998, Rochester was known as a city. However, due to an administration 'hiccup' when the new authority, Medway was formed, the sitting local government forgot to protect its city status. Now it would take a King's charter to reinstate it as a city, so for now, it's Rochester town.

Rochester, from its early inception, has developed from a small Saxon village into one of England's best-known historic cities.

In 43 AD, Emperor Claudius was in charge back in Rome. His army under the command of Aulus Plautius Silvanus started the Roman Invasion of England by attacking Rochester. By all accounts, it was a bloody battle and raged for over two days.

Over the years, this conflict is now known as the Battle of Medway.

However, its actual location (so it is believed) was at a crossing of the River Medway, close to Rochester Town itself.

Now before anyone screams. "It's a city! It's a city! Not a town" it should be remembered that it was only granted city status by Henry III via a Royal Charter in 1227.

However, it was not the Romans who left England in the third century or the Anglo-Saxons, who built a stone castle at Rochester, as this did not happen until 1088 after the Norman invasion.

More of the Castle and other history titbits later. For now, let's talk Christmas!

So let's start with the 'Man who invented Christmas'! (A great film by the way, and one to watch to get you in the festive swing of things).

Charles Dickens

- Charles Dickens (middles names John Huffam) was born on Friday, 7 February 1812 (a significant date for later on!), in Portsmouth, England and died on Thursday, 9 June 1870 while living at Gad's Hill, near Chatham, Kent, he was only 58 years and four months!
- Spooky!: My favourite Dickens Character 'Ebenezer Scrooge' was given the same birthday (except year) as Charles himself; 7 February 1786, which was a Tuesday, and he died on Tuesday, 19 December 1843 aged 57 years and 10 months. Note their age, when they both died they were less than six months apart! Now that's Spooky!
- Incidentally, my birthday is also on 7 February. Maybe that is why I love Christmas so much.
- Born in Portsmouth, his family moved to Chatham when Charles was aged 5. They stayed there, and Charles started his education at William Giles

School. In 1882, his dad transferred to London and the family re-site in Camden Town. (Looking back, maybe not a great move for the family!)

- In 1824, John Dickens was sent to the debtors' prison (Marshalsea Prison, Southwark), and young Charles was dispatched to take work at Warren's Blacking Factory. Remember he is still only 12 years old!

- Charles ended back up in Medway and died in Gads Hill, near Chatham. But this is not where the intrigue ends! His journey to his final resting place reads like a plot from one of his novels.

- "A Victorian gentleman's wishes are defied by an interfering and ambitious cleric, leaving a grave without a body in the hallowed Lady Chapel of Rochester   Cathedral."
https://www.bbc.co.uk/news/uk-england-kent-16834429

- Even more spooky, his final demise and resting place almost pay homage to his last work (unfinished), 'The Mystery of Edwin Drood'. The plot revolves around the possibility of a corpse hidden in Rochester Cathedral. So what happened?

- Charles Dicken's remains seemed to have been the source of a dispute between Rochester Cathedral and Westminster Abbey. One (Rochester) believed Charles' wishes, included being buried in the Cathedral and facilitating this after he passed away. As the graveyard was full they proceeded to dig a vault inside the Cathedral. Yet at the same time then, Arthur Stanley (Dean of Westminster) was looking to bolster the kudos of 'Poets Corner', and who better than to have Charles Dickens laid to rest within the Abbey, so he was whisked away and buried in Poets Corner and the Dean and Chapel were left with an empty vault in Rochester Cathedral!

- What is a matter of record is that on a Monday morning, a vault was formed in the Cathedrals St Mary's Chapel. However, by Monday afternoon, a decision was made that he should rest in Westminster Abbey, and arrangements were made to transport his remains there. Then on Tuesday morning, Rochester Cathedral's empty vault was filled, and the floor stones were reinstated.

- A nice local touch though, ends this tale, and that is that the tolling of the Cathedrals Bell rang out while the vault was filled.

- Many of Charles Dicken's novels made reference to Rochester and the local area, and even today, the Christmas Dickensian festival is held in his honour.

So what about Christmas markets? In Rochester, there is an abundance of Christmas festivities including the Christmas market in the Castle and the **Dickensian Christmas Festival** (*https://www.rochesterdickensfestival.org.uk/news.htm*) around the town.

Dickensian Christmas

The site above will provide you with all you need to know about the Christmas Festival, but needless to say, the atmosphere, the colours, the costumes, the drinking, the games etc., all set in quaint Rochester, is a perfect backdrop to start your Christmas preparations. If you can watch The Man Who Invented Christmas before your trip, it may even whet your appetite further.

So, as Rochester reverts back to the Victorian era with the Dickensian Christmas, the emphasis is on the Christmas novel 'A Christmas Carol'.

As you would expect, or maybe just in your mind's eye (which always sees through rose-tinted glasses, not beer goggles!), it always snowed at Christmas in Victorian times.

In Rochester, your memories/expectations are fulfilled as it always snows here at Christmas time, albeit it may with the help of an artificial snow machine, in case the real stuff doesn't make an appearance.

Your senses are further treated by the aroma of 'Chestnuts roasting on an open fire', the sounds of skates cutting across the ice, and the warmth of a mulled wine to soothe the inner chills of the winter air.

The festival culminates with the Dickensian Candlelight Parade, which meanders through the streets ending up in the shadow of Rochester Cathedral.

Check the website above for the dates of this event, which generally runs over a weekend leading up to Christmas. You may be lucky to tie it in with the Christmas market visit. Bear in mind pubs may be busy.

Christmas Market: Rochester Castle

Set within the grounds of Rochester Castle (check out the opening times and dates from the official website), this is a unique setting for a Christmas market. Although this may not be the biggest Christmas market, it is well worth a visit, especially if you can tie in your visit with the Dickensian festival as well. But, let me have a quick word in your ear. Remember, Rochester is a small town with small roads, so you need to plan your visit well ahead of time. If you are driving for the day, make sure you schedule your visit to avoid rush hour etc.

Another point worth remembering is that the Castle grounds are not vast. And as the Christmas market is only on intermittently (weekends etc.), it can get jam-packed.

That said, so can every Christmas market (Valkenburg in The Netherlands should definitely be visited on weekdays if you don't want to feel claustrophobic, and not at weekends).

So the old adage 'proper planning prevents piss poor performance' has never been more appropriate.

Don't be put off, but don't expect an empty market.

So a little bit about Rochester Castle.

- Rochester Castle was built here as it is a strategic location.

  o Located in Kent, South East of England. This area was a prime site used to invade England throughout the ages.

  o It was soon recognised that placing this Royal Castle here would serve as the first line of defence to protect England from invaders.

  o It is also the second oldest Cathedral in England, having been built in 604 AD. The oldest one is Canterbury Cathedral which was set up in 597 AD.

- Built along Watling Street, a Roman route which led from the SE Coast through Rochester up to Viroconium Cornoviorum, now occupied by Wroxeter, Shropshire.

- The Bishop of Rochester Gundalf (1024–1108) [Not Gandalf☺] constructed the first stone castle where it now stands. Some of the original stonework can still be found along the Western Curtain Wall. Gundalf also had an input in Rochester Cathedral and the White Tower in the Tower of London.

- In 1127, Henry I gifted the then Bishop (William De Corbeil) with the Castle and gave permission to build a stone keep within the Castle. This is now one of the best preserved and oldest keeps in England.

- Another upgrade happened early in the 1200s when King John feared an attack from the French.
- King John was on the throne of England for 17 years from 1199.
- Some believe he usurped the throne from Richard the Lion Heart and through his antics, poor leadership, and general poor kingship, has since gone down as one of the worst of Kings of England.
- John's reputation was damaged further after losing loads of land in France, which nearly bankrupted England. This upset the Barons and was one of the reasons why they rebelled. This eventually led to them forcing King John to sign the Magna Carta in 1215 CE. (Which had the effect of curbing the powers of the Crown).
- Legend has it that he was also plagued with a notorious outlaw Kevin Costner... sorry Robin Hood, but there is very little evidence this person existed (Robin, not Kevin), but why let that ruin a great story.
- The Irony of ironies; King John tries to take Rochester Castle!

  o In 1215 (October time), King John laid siege to Rochester Castle to try and grab it back from the barons.
  o His army successfully took Rochester Bridge (and tore it down to prevent any attack from

his rear), then went on to capture the Castle's Bailey.

- o He then used siege engines to bombard the Barons (and others), inside the keep, with stones.
- o Having little effect, King John changed tactics and instructed his miners to tunnel underneath the southeast tower.
- o Using fire (fuelled by pig fat), he set the foundations alight, which quickly made the tower unstable, causing its collapse.
- o Finally, due to starvation, the Barons surrendered, but his victory was short-lived as King John died of dysentery in 1216.

- It was not until Henry III came to the throne that the fallen tower of Rochester Castle was rebuilt.
- Over the centuries, the Castle had other sieges and battles but eventually stumbled into general decay, so much so that Samuel Pepys wrote about the poor state of Rochester Castle in one of his diaries:

"I did there walk to visit the old castle ruins, which hath been a noble place, and there going up, I did upon the stairs overtake three pretty maids or women and took them up with me, and I did besarlas muchas vezes et tocar leaur mains (roughly translated as I kissed them on their patches [beauty-spots] and touched their hands and necks to my great pleasure] and necks, to my great pleasure); but Lord, to see what a dreadful thing it is to look down precipices, for it did fright me mightily and hinder me of much pleasure which I would have made to myself in the company of these three if it had not been for that."

- Even Charles Dickens refers to Rochester Castle's poor state.
- In two of his books: The Mystery of Edwin Drood and The Pickwick Papers, he uses terms such as 'glorious pile, frowning wall—tottering arches—dark nooks and crumbling stones' all of which sets the scenes perfectly for the storylines.

Ghost Alert! The Lady in White

Let me introduce you to Lady Blanche de Warrene, a rare beauty who caused quite a stir amongst the men. Now the story which follows could have been a great plot for a Shakespearian Tragedy and a triangle which knocks Arthur Guinevere and Lancelot into the realms of a minor triste.

So let's set the scene:

- Team A: Rochester Castle; forces loyal to Henry III (who wanted more funds from the Barons for his war chest... where have we heard that before?). These were led by John de Warrene, who had made the Castle his home, and a crusader Sir Ralph de Capo (current suitor to Blanche Warren, daughter of John).
- Team B: The Barons; decided to re-take Rochester Castle from the King and thereby take control of the bridge and control access along the river. The army was led by Simon de Montfort, and in his ranks had a knight called Gilbert de Clare, seventh Earl of

Gloucester and rejected suitor of Lady Blanche! I think he never really got over her rejection.

- So the siege starts, and despite the best efforts of the Barons, the Castle remains resolute in its defence, and little progress is made.

- Now, Simon de Montfort is stuck on the wrong side of the river, but he devised a cunning plan: "as cunning as a fox who's just been appointed Professor of Cunning at Oxford University?" (from Blackadder's most cunning quotes and one-liners).

- Getting Gilbert to take a company of men to Aylesford, where they would cross the river and sneak back, flanking the Castle, ready to pounce when required.

- On Good Friday morning, Simon de Montfort played his bluff card. He had set fire to the bridge; unsurprisingly this caused a cloud of thick black smoke to rise up and engulf the Castle. This really upset the Castle's residents (a lot remember what another fire did at an earlier siege!). Enraged, they summoned even greater determination and redoubled their efforts to 'see off' De Montfort.

- As planned, De Montfort feigning fear, pretended to retreat.

- Warrene's men became cocksure after their perceived success. This, coupled with the rumour that the King's own forces were on their way to join in the foray, raced from the safety of the Castle and gave chase… have you guessed it yet?

- Leading the chase was Sir Ralph.

- The Castle was now vulnerable, so Gilbert and his men attacked from the flank.
- Lady Blanche de Warrene was watching the battle from the southern wall of the Castle and was recognised by Gilbert. Here the story gets a bit greyer.
- The crux of the matter was that Gilbert, wearing the same gear as Ralph rode unchallenged into the castle and raced up to the battlements and grabbed Lady Blanche.
- Imagine her surprise when she saw it was her EX, rather than her future husband, who had come to her rescue!
- As they struggled up on the battlements, can you guess who saw them?
- Sir Ralph (an excellent archer) grabbed a bow and launched an arrow high in the air. It would have been a great shot, had it not deflected off of Gilbert's armour and into the chest of Lady Blanche, killing her almost instantly!
- That night a spectre dressed in white with flowing black hair was seen floating along the wall with an arrow protruding from its chest.
- Legend has it that she walks that lonely path every Good Friday night.

An anonymous poem from 1829 also recounts the tragic love triangle.

Mark well the southern battlement, for there Earl Warren's lovely daughter, Blanche the fair, Fell like a flow'r, which is by some rash hand cut down, before its beauties can expand. Ann arrow 'twas, from noble Capo's bow, That pierced her heart and laid the maiden low And it is said—how true, we ne'er can tell. Ever at midnight, when the Priory bell Proclaim'd that boding hour, the maid was seen against the fatal battlement to lean!

'Another of Geoff Rambler's City Ramblings. March 2017'

So let's go for a drink.

Route: Leave the Christmas market via the Bakers Walk entrance and turn left up the hill, carry on into Boley Hill, then turn right into St Margaret's Street. Your starting point is just down here on your right.

Christmas Factoid:

According to data gathered from Facebook, two weeks before Christmas is one of the two most popular times for couples to break up; one would guess to avoid buying presents. Christmas Day is the least popular.

The Coopers Arms

Address: 10 St Margaret's St, Rochester ME1 1TL

Hours:

Monday	—	12:00 pm–11:00 pm
Tuesday	—	12:00 pm–11:00 pm
Wednesday	—	12:00 pm–11:00 pm
Thursday	—	12:00 pm–11:00 pm
Friday	—	12:00 pm–12:00 am
Saturday	—	12:00 pm–12:00 am
Sunday	—	12:00 am–11:00 pm

The Coopers Arms was built during the reign of King Richard the Lionheart (1189–1199).

The first known inhabitants of this house (soon to be an Inn) were the monks from the nearby priory St Andrews. These monks were renowned for their skills and aptitude in brewing ales and wine. But more challenging times were ahead, and it fell into disrepair during Henry VIII's Dissolution of the Monasteries.

It finally opened fully as an Inn in 1543 and has been a resolute part of the city serving fine cask ales/beers ever since. No matter the structural changes that happened over the centuries, it still holds on to its original charm.

We start in a lovely traditional English pub exuding atmosphere, enhanced by an open fire when called for. Cask ales are on sale, along with beautiful food, so sit, relax, and soak up all the joie de vivre while contemplating that there is more of this just around the corner.

Ghost Alert (Already!)

So it is believed that this pub is haunted by a spectral monk who makes an annual appearance in the winter month of November. This poor fellow is said to have committed such a grave sin against the Brethren of Coopers (it has been suggested it may have involved an illicit affair with a local woman, but this is only anecdotal) he was killed by burying him alive behind the walls of this very pub.

He has been seen emerging from his walled tomb, drifting and surveying the bar, before returning to his grave within the wall. People who have seen this apparition say that his face appeared angry, which is hardly surprising in the circumstances.

There are even some reports of another ghost sharing this abode. Although not prolific, reports of an old fellow with 'wild eyes' show up occasionally!

Whatever the anecdote/legend or myth, there is something that has happened to people here in the past, which asks more questions than provides answers. So if you are visiting on a cold winter's night in November. Personally, I would have one eye on my beer and one eye on the walls.

Route: Come out of the pub and turn left along Love Lane at the T junction. Turn right into St Margaret's Street. Continue along here, then turn left into Vines Street (Kings School Rochester is on the corner as you turn left). Carry on up Vines Lane until it ends. You then take the second exit into Maidstone Road. Your next watering hole is just down this road on your right-hand side.

While on your route, here are some titbits of information to whet your appetite.

Ghost Alert

- As you enter Love Lane, be careful you don't bump into half a lady!
- There is a record of a local resident who lived in the town in the 1950s, was walking his Great Dane dog and turned into Love Lane. At the time, there was a row of derelict houses along this road, but these were bombed and destroyed during World War II.
- As they approached one of the derelict houses, the dog became very agitated and made it clear he did not want to go any further.
- The gentleman peered into the ruins and saw a lady wearing a white blouse, but on closer inspection, there was nothing below her waist and she seemed to be floating on air!

- At this point, the account stops. Obviously, he did not hang around to check out the apparition further.

King's School, Rochester (Bridesmaid Again!)

This is the second oldest school in the world, apparently! Can you guess who is number 1? Canterbury again.

- Set up in 604 Anno Domini, and it is still very relevant today, as it welcomes around 700 pupils annually.

As we continue up Vines Lane, and before we turn right into Maidstone Road, on the left is Crow Lane, and just down there is a splendid piece of historical architecture:

Restoration House

King Charles II apparently stayed there:

- Ahead of the Restoration of the monarchy in England in 1660. Which marked the return of Charles II as King following Oliver Cromwell's Commonwealth.
- King Charles II stayed in Rochester the day before the ceremony. Using this as his overnighter base on his way to claim the throne in London.
- The house he stayed in was not called Restoration House at the time and was renamed to commemorate the event.
- What is actually there today? A conjunction of two medieval buildings, combined in either the late sixteenth or early seventeenth century.

- The medieval property is right in the centre of Rochester in Crow Lane.

Now, what has an Emu and comedian got to do with this grand old building?

In fact, Rod Hull bought it in 1986 for £270000 hoping to save it from being knocked down and built into some concrete monstrosity.

Mr Hull, then set about restoring the property. However, after spending nearly half a million pounds on it, he went into bankruptcy.

It was taken over by the receiver in order to pay his tax bill.

In reality, everyone owes Rod Hull a great debt of gratitude. If not for him, in all likelihood, it would not be here today.

Other guests to the house apparently include Samuel Pepys and Charles Darwin. However, I doubt at the same time, unless Samuel came back as a spectre, talking of ghosts! (What a Segway!)

Ghost Alert

Now it is rumoured that in the room where he slept on his way to London, a shadowy figure of what appears to be a man is drifting across the room from the area of the bed and through the bedroom wall.

A more detailed sighting is the legend of a young lady ghost, who is thought to be either;

- The spirit of a housemaid, who fell for the charms of her boss and became pregnant, and walks the house in despair.
- Or it is not a maid at all, but a beautiful lady dressed in white simply trying to escape from the house through the front door.

However, an article in Scribners Magazine 1893 Volume XIII refers to the ghost as a nun carrying a baby child. There are even reports of a noose around the ghost's neck, but I can neither confirm nor deny that this is true.

Like a trick in the tail, other legends say that the nun/lady is followed closely by a monk. The mind is now racing as to what this really means, but who knows?

And on that thought, let's go to the next pub, just down the road.

The Granville

Did you know?

The longest-running Santa Claus parade happens in Illinois. Plenty of communities hold Santa Claus parades, and each one is special in its own way. However, the parade that happens each year in Peoria, Illinois, is the oldest parade of its kind in the United States. It's taken place every holiday season since 1888—that's well over 100 years of festive fun.

Address: 83 Maidstone Rd, Rochester ME1 1RL
Hours:

Monday	—	4:00 pm–11:00 pm
Tuesday	—	4:00 pm–11:00 pm
Wednesday	—	4:00 pm–11:00 pm
Thursday	—	4:00 pm–11:00 pm
Friday	—	4:00 pm–12:00 am
Saturday	—	12:00 pm–12:00 am
Sunday	—	12:00 pm–11:00 pm

Smallish pub with a local community following. Original external Meux Brewery tiling on the front wall pillar. Its marketing strategies may be a little off the mark, advertising 'normal beers and average wines'!

This pub has been around for over 100 years and still looks good.

Route: Come out of the pub and turn up King Street, then take second left into Cazeneuve St At the following crossroads, go straight across into Victoria Street, and stay on here until you get to a T junction with High Street. Turn right, and here is where you need a will of iron. Stay moving down the High Street until you reach The Nags Head (no stopping off on the way as we will be coming back this way!)

Okay, as we walk, let's think about pub names and who started the idea of naming them.

Have you ever considered why we call pubs, The Queens Head or the Kings Arms or Ship and Shovel instead of Beer House or Ale Place? Well, it all started many years ago...

- Before we talk about the great English/British pub, let's go back a bit further in time, back to the Romans.
- On my trip to Pompeii, you are told of all the bars and restaurants they have uncovered, and even a brothel. But how did they know what was what? Well, they had signs showing wine carriers, some Roman bars had vine leaves... signalling wine etc.! I will let you guess what imagery they had to signify the brothel!
- But when the Roman Empire expanded into England, vines were in short supply, so they used branches of bushes to denote a pub. This may be why we have pubs today with names like 'The Old Bush' 'The Mulberry Bush' etc.
- Now jump forward to the twelve century in England.

- As the popularity of drinking houses grew, proprietors of the pub wanted to draw customers into their establishment and to keep them coming back. So to stand out from other pubs, they needed a distinguishing feature. This led to them hanging specific items outside, like bells, shapes of stars, anchors etc., which led to names such as The Three Bells, The Star and The Anchor.

- Rather than just objects, some others chose to refer to issues of the day, i.e. The Crusader or Pilgrim's Rest. But what must be remembered at this time, writing words would have been impractical as only a few people could read, so the birth of the pictorial pub sign started around this time. This was further supplemented when in 1393, King Richard II passed an Act making it a requirement in law for drinking establishments to advertise to the public that they were indeed pubs.

- Strange but true: Do you know where the pub name The White Hart came from? Spookily the White Hart was Richard II's personal emblem.

Now you may not know, if you are new to the area, that this part of town is known as Troy Town.

Troy Town

So why the Troy Town?

One explanation could be found if we go back in history.

Have you ever heard of Turf Mazes?

- A turf maze is a puzzle, made by creating a path in a level area of short grass or turf. These were affectionately known as Troy Town or similar (the reference to Troy probably has something to do with folklore, which suggests the city walls of ancient Troy were constructed in such a way, any invading army who entered them became confused and unable to find their way out). It is widely believed that it is a turf maze Titania is speaking about in Shakespeare's *A Midsummer Night's Dream* where she says:

The nine men's morris is fill'd up with mud;
and the quaint mazes in the wanton green,
for lack of tread are undistinguishable
A Midsummer Night's Dream, Act 2, Scene 1
Translation: LitCharts

- Nine men's Morris is basically a game like noughts and crosses played on the ground so that the quaint mazes had been made unrecognisable by the effects of a flood and covered in mud. We assume it is a turf maze she was referring to when looking at the pathway, as tall hedgerow mazes did not exist in Shakespeare's time, outside of Palaces etc.
- So there was a maze here many many moons ago.

On the Corner of Victoria Street and East Row

If you can picture the scene, a few hundred years ago this area would have been suffused with the aromas associated with The Troy Town Brewery.

Why build a brewery here? Well, it may have had something to do with a spring supplying fresh water readily available on the site.

Now although the beer has been long gone, if you look to the top of the building, you can see etched into the stonework 'Woodhams and Co Ltd' 'Established 1750'.

Now this old Brewery was initially set up by Henry Shepherd and ran for over one and half centuries before closing its doors in 1919.

The Brewery was modernised around 1860 after which it traded as the Troy Town Steam Brewery.

Even back then, advertising and being part of the local community was an essential part of the business model so that Woodham's and Levy brewers registered a barge called 'Fanny' and competed in an 1874 Medway sailing match.

By the end of the century, after the ownership shifted backwards and forwards amongst the families, it finally became a Limited Company named Woodham's and Co Ltd.

Even at its height of production, and even when it turned electric, it only ever employed a maximum of 25 people.

In 1913, the Brewery was updated with the introduction of electric power.

Finally, it closed its brewing doors in 1918.

Ghost Alert!

Surprise, surprise, we may have another ghostly occurrence, but this one may be able to be explained.

In April 1893, the owner of Restoration House, Stephen Averling Esq, an exemplary upstanding community member, relates an eerie tale of a ghostly fright which happened to him at Restoration House.

In the dead of the night, strange noises could be heard, with at first glance had no explanation, but on further inspection, the noises were deemed to be from a disused water tank. The story goes that although the water tank was no longer supplying the house, it used to be fed from the brewery supply, and the noises travelled through the empty pipes as the pumps were brewing at night... So they say!

As we approach Star Hill, here is something you may not know.

St Catherine's Hospital

- In 1316, Symond Potyn, Master at the Crown Inn (just down by the bridge, and although not on our crawl, it's well worth a visit), Rochester, left money to fund the Hospital of St Catherine.
- This was there to provide people suffering from leprosy and other such diseases care and treatment. Originally built at the bottom of Star Hill in the High Street.
- When it was first built, it was on the outskirts of the city.
- This was to protect the community from what was considered to be infectious diseases.

As one can imagine, in those days of yore, there were bound to be some crazy rules!

The Rules

- No resident to be absent from 'the hospital after the sun goes down'.
- No resident to haunt the taverns. If a drink was wanted, it was to be brought into the hospital.
- No resident to be argumentative or use ribald language.
- At a certain hour of the morning and evening, the residents were to pray for the Sovereign and all the realm of England and for Simon Potyn while alive and for his soul when dead.
- If any resident should cause trouble in the city, the Vicar of St Nicholas and twelve good men of Eastgate were to complain to the Prior and the resident shall be put out of the hospital forever without them taking anything but their clothing.

The original hospital was eventually knocked down and then rebuilt on the apex of the hill, and the reason given for this re-siting was that was a bit more 'airy' up there.

So gird your loins and summon up all of your will as we yomp down the High Street until we reach the Nags Head on your right. We will talk some more about the High Street on our way back down!

The Nags Head

Address: 292 High Street, Rochester

Hours:

Monday	—	11:00 am–11:30 pm
Tuesday	—	11:00 am–11:30 pm
Wednesday	—	11:00 am–10:30 pm
Thursday	—	11:00 am–10:30 pm
Friday	—	11:00 am–12:30 am
Saturday	—	11:00 am–12:30 am
Sunday	—	12:00 am–11:00 pm

No one is sure why the pub was called the Nags Head. Could it be because of the road name or because there was a police station and stables on the site before the pub? Unfortunately, there is no evidence either way.

This 400-year-old pub (renovated in the nineteenth century) has a weatherboard frontage. This makes it a quite picturesque, old English style pub. Also, the interior décor does not disappoint, and many of the adornments hark back to a different era.

Ghost Alert

So how is the Nags Head haunted? Well:

- A 'lady of the night' Aggie was being held on charges, which one can only assume to be around her profession in the Police Cells, which were at the time at the back of the pub (before it was built).
- Not being very happy with the situation, be that her life or the charges and likely sentence, Aggie hanged herself in the cells.
- The public and staff attest to seeing the ghostly lady on the stairs of the cellar... what do you think was there before the cellar?
- Previous Landladies and landlords have at different times heard crying in the night, as well as clanking and banging!
- It is also reputed that the area just outside the pub was known as Hangman Lane (one because the hangman lived nearby, and two as local ne'er-do-wells were hanged here).

Route: Okay, turn left when you get out of the pub and walk back up the High Street, going back the way you came down, carry on this road until you reach the Queen Charlotte on your right.

So as you wander down the High Street, and if you are lucky enough to do this in and around April/May, you may coincide with the local celebrations of The Sweeps Festival. Considered by some as being the biggest folklore celebration in England. Here is a potted version of The Sweeps Festival;

- In days of yore, Sweeps only ever had one day off in a year, and this was May Day.
- May Day is an age-old European festival which is to say goodbye to spring and to say hello to the start of summer.
- Generally centred on 1 May, which is halfway between the vernal equinox and the summer solstice.
- NB: An equinox is the time of the year when the daylight and night-time hours are about equal in length. The vernal equinox (March) marks the start of spring, while the autumnal equinox (September) marks the beginning of autumn.
- Solstices are the two times of the year when there are either the most daylight hours (Summer Solstice in June) or the most hours of darkness (Winter Solstice in December).
- Solstices are considered to mark the start of summer and winter.
- Note these dates will be reversed in Sothern Hemisphere.

- Back to May Day. In the past, these celebrations sometimes started the night before on May's Eve, which makes sense. Why waste a night?
- Traditions often included celebrations of nature coming out of winter by gathering wildflowers and green branches, weaving floral garlands, crowning a May Queen and setting up a Maypole.
- Back then, the consensus was, that the Maypole was a living tree.
- But as the years passed, it morphed into a tree trunk and eventually a pole.
- The erect May Pole is believed by some to exemplify man's energy (it is about new growth after all). But having a naked pole is no good to anyone, so the ribbons and floral garlands that are wrapped around it, represents the female energy!
- Whatever the hidden meanings are, the Maypole is a splendid spectacle showing that spring has sprung and Mother Nature is now in the mood for growing.
- As you can imagine, this was when Sweeps wanted to kick up their heels and have fun!
- This led to morphing several folk traditions into one event, which meant gallivanting through the streets, dancing, singing and letting their hair down. (Morris Dancing will be covered in a while!)
- Now comes the Jack in the Green.
- The origins of this were first recorded in the seventeenth century. This originally involved milkmaids decorating their dresses and pails with flowers for the celebration. Samuel Pepys recorded in

his diary that while watching the London May Day parade of 1667, he noted milkmaids had 'garlands upon their pails' and were dancing behind a man playing the fiddle.

- Another account a year later saw the milk-maids had abandoned their pails for a silver platter balanced on their heads. On the platter were various objects built into the shape of a pyramid adorned with flowers and ribbons etc.

- Still following fiddlers (or even a bagpipe player), they went around the town accepting occasional payment (in some form or another). These milk-maids were accompanied by musicians playing either the fiddle or bagpipes and went door to door, dancing for the residents, who gave them payment of some form.

- Jack in the Green; the Green man appears in many cultural 'Spring' festivals and is linked by many fertility (here we go again with the MAYPOLE!).

- Other Countries' festivals include Walpurgis Night (Germany): Beltane (Eire) and Calan Mai (Wales).

- Okay, The Sweeps Festival revival came about after it had lain dormant for a long time.

- Believe it or not, the law and literature actually worked together, albeit independently, to affect this festival.

- In 1840, an amendment was made to Chimney Sweeps Act.

- The change raised the age of children climbing up chimneys to 16.

- Sadly, without any enforcement instruments it was largely ignored, and kids of 10 and younger were still made to climb!
- Next, in 1863, Charles Kinsley's story about kids being forced into climbing chimneys (The Water Babies) raised the issue in the public's mindset.
- 1864 Parliament again tried to amend the Act but failed to make an impact.
- Finally, in 1875, an amendment to the Act required Chimney Sweeps to be licensed.
- The Act now made it clear that the Police would arrest offenders and charge offenders under The Act.
- So with these changes, the number of Sweeps began to dwindle so too did the festival, with Rochester seeing its last at the start of the twentieth century.
- In the 1980s, one man, Gordon Newton (historian/musician), was paramount in the resurrection of the festival.
- This celebration has since grown continually, and it now attracts a wide range of visitors, some musicians, some dancers, some come to join in the parade, and lots of spectators (it may have something to do with the pubs and alcohol).

Something Strange for Postman Pat in Rochester

- In Rochester, one of the oddities is the 'out of the normal' post boxes dotted around.
- The first one we come across on our crawl is outside, as you may expect, the Post Office on the High Street.

Just before we reach the Queen Charlotte at the southern end of the High Street. Nothing particularly amazing about this, but it's an old oddity. Can you spot it? It has the following written on it. 'Posting Box'.

More later, now let's pop into the QC for a drink.

Christmas Quote:

"The best way to spread Christmas cheer is singing loud for all to hear."

– Will Ferrell *Elf*

The Queen Charlotte

Address: 159 High St, ME1 1EH Hours:
Hours:

Monday	—	11:00 am–11:00 pm
Tuesday	—	11:00 am–11:00 pm
Wednesday	—	11:00 am–11:00 pm
Thursday	—	11:00 am–11:00 pm
Friday	—	11:00 am–1:00 am
Saturday	—	11:00 am–1:00 am
Sunday	—	11:00 am–11:00 pm

This old pub was known as the Royal Charlotte in a previous incarnation.

This drinking establishment replaced another pub, which was on this site many years ago. This newer version harks back to the early 1800s.

The name comes from either the

A.

- 'Queen Charlotte' was the consort of George III (The Mad King!).
- Not many people know about Charlotte, but she was the lady who founded Kew Gardens.
- She had 15 kids (not a lot to do at night back then).
- Charlotte was also an Arts patron.
- She even knew Mozart, who, at eight years old, performed for her. Due to the success of this performance, he was invited back to play again on the occasion of George III's anniversary of his accession to the throne in 1764.
- Charlotte obviously made an impression on young Wolfgang, as he dedicated his Opus 3 to her when it was published on 18 January 1765.
- In the USA, they refer to her as the first 'Black Queen'.

B.

- Or was it named after a ship? No one knows for sure.

Ghost Alert and Factoid

- The Queen Charlotte is not content to have just one ghost. They have two and one is a humdinger.
- First off, reports from the staff have told of glasses being smashed for no reason in the bar area, and the aroma of Lavender is often encountered in the cellar. Both these lead to the assumption that an old lady is frequenting the pub. Although, no actual sightings have been recorded to determine what she looks like.
- Now the second ghost is a little bit more of a character, as he was in real life. Mr Frederick Adolphus Gould was the landlord from 1908, but what was not known at the time was that it was an alias!
- In fact, his real name was Adolphus Schroeder a German National.
- Let's think, what was happening in the world back then? Not long before WW1
- So after giving up the pub in 1913, he moved to London and worked as Merchant selling tobacco/cigars.
- Now what silly billy would leave incriminating evidence behind when moving out of a pub... erm!
- You guessed it, he did! The new landlord found a secret stash of old documents, which included maps and other Admiralty information, as well as a letter

which alluded to the fact that Mr Schroeder was the holder of the Iron Cross, which he was given for his actions in 1870 during Franco-German war.

- It also said that he then worked for the German Secret Service.
- Now coming under suspicion, he was put under surveillance and was arrested and convicted of espionage and sentenced to six years of hard labour in 1914, followed by deportation. No real evidence has been found as to what happened to him after his sentence was complete.
- That is until staff reported strange noises coming from the attic, but on inspection, nothing was there. However, the noises continued to be heard by different individuals, who believed it was Gould trying to get in contact with his German counterparts.

Also, what some people overlook, is that in 1762 the King and Queen moved from their house in St James Palace into an unassuming building just down the road known as Buckingham House. From then onwards, it would be known as Buckingham Palace and Queen Charlotte loved it there.

In 1792, Charlotte also bought a country retreat in Windsor known as Frogmore House in Windsor:

Route: Come out of the pub and turn right down the High Street, and continue to walk down (passing other pubs, unless you're feeling sober and want to jump in and out) the High Street until you reach The Eagle on your left.

As you go, behind the buildings on your left are the Cathedral Gardens, which also has some history.

- Let's talk about a well-known character who was the Bishop of Rochester that is Bishop John Fisher, 1469–1535.
- He was a Chancellor of Cambridge University and Rochester's Bishop.
- His big mistake (well depends on where you sit on religious conviction) was that he did not support Henry VIII's divorce!
- To make matters worse, he did not take kindly to Henry's self-declaration as Supreme Head of the Church of England.
- Back in those days, there was no community service, and being found guilty of treason, he was made to take the long walk to meet the axeman.
- Executed at the Tower of London on 22 June 1535.
- Before he left this mortal coil, he made a short statement giving thanks, for facing death without fear.
- Another one not to hold a grudge, Henry ordered that his body be left on display all day. He then had his clothes ripped from him and dumped into a newly dug grave at All Hallows by the Tower Church graveyard.
- He was later reburied in the Chapel of Saint Peter. (headless).
- To ensure things were not forgotten the 'tradition' of the day was to have the heads of traitors etc., mounted on top of a spike on London Bridge.
- John's head apparently only stayed for a few days before being replaced by Sir Thomas More's. The

rumour is John's head was just thrown into the Thames.

- That may seem an excellent theme for a Ghost Alert, but no! It seems he was content to pass over to the afterlife. After all, he was a martyr.

History of the Cathedral Gardens

- Back in the fifth century, King Æthelberht not only gave the land for the Cathedral but also gave part of the old city (which was walled), on which the gardens now stand.
- The area was added to and expanded in the eighth and ninth centuries.
- Over the centuries, buildings were added, demolished and rebuilt. However, back in the 1930s, 'Perpendal House' was removed and replaced with the Gardens. (which were then opened to the public).

The Vines

- Now, ensconced in Kent's history is Billy the Bastard (William the Conqueror, whose coronation was 25 December 1066: see the Christmas Segway!).
- Well, his half-brother, Bishop Odo, supposedly also gave land to the monks. This time though, it was outside of the city walls and was to be used to expand the gardens further.
- This area became known as the Vines. (Remember earlier).

- Now monks are generally associated with brewing beer (especially in Belgium!), but not renowned for wine.

- However, there is a reference to their expertise in this area as well in the Registrum Roffense (1319–1352), which is basically a collection of ancient records which recorded the history of Rochester Cathedral and states there indeed was wine growing and wine-making facilities here.

- John Worlidge wrote later in his '*Vinetum Britannecum (1767)*' that: "Great quantities of grapes grew here, and produced fine wines; Bishop Haymo de Hethe presented King Edward II (who was then at Bockinfold) with a taste of his wine (C1340)."

Vinefields and King's Orchard—Rochester Cathedral.

- So as a homage to this piece of history, they named the gardens near the King's Orchard, 'The Vines'.

Christmas Quote: (Probably the worse one ever!)

"Nothing says holidays, like a cheese log."

– Ellen DeGeneres

The Eagle Tavern

Address: 124 High St, ME1 1JT

Hours:

Monday	—	12:00 am–7:00 pm
Tuesday	—	12:00 am–11:00 pm
Wednesday	—	12:00 pm–12:00 am
Thursday	—	12:00 pm–12:00 am
Friday	—	12:00 pm–12:00 am
Saturday	—	12:00 pm–12:00 am
Sunday	—	12:00 am–7:00 pm

The Eagle now describes itself as Rochester's premier music venue.

This small pub has a large garden at the rear where you can still see the old city walls or what remains of them. I cannot determine with any hard evidence why the name The

Eagle was used, but it is suggested it may have been in homage to the Emblem of St John the Evangelist.

What is known is that The Eagle has been here for quite a while, indeed in the Kentish Gazette on the 4 September 1849. The headline read;

Rochester, Impudent Robbery

On Wednesday last Mr Grist, of the 'Eagle Tavern', Rochester, was robbed of his cash box, containing £15, by a man having the appearance of a respectable traveller, who had sojourned there for the night. In the morning, he took his departure, and shortly afterwards the robbery was discovered. It appears that he got possession of the box, which was left in a drawer locked up in Mr Grist's bedroom, by means of a skeleton key. He was unknown, and Grist, being ignorant of the road he took, was unable to follow him.

http://www.dover-kent.com/2014-project/Eagle-Tavern-Rochester.html

Route: A short hop to the Two Brewers. Turn left out of the pub and crawl down the High Street until you see the Two Brewers on your right.

As it's only a few yards, just a titbit to get you there.

As we move just down the road, let's dig a bit deeper into the Morris Dancing we spoke about earlier:

Morris Dancing

- This is an old English folk dance with the dancers wearing bells and using sticks and handkerchiefs to interact with each other.
- It is probably a hybrid of the dances which were performed in the European courts back in the 1400s. This form of entertainment was known by names such as 'morisco' or 'moreys daunce'.
- They were performed as solos or a 'circle' surrounding a person with a view to gaining their favours.
- The first written record of Morris Dancing in England was actually a payment record, paid by the Goldsmiths Company for a London-based event, which attributed the cost of seven shillings.
- From this point onwards, Morris Dancing keeps popping up in records.
- In fact, by the 1500s, Morris Dancing had embedded itself in church celebrations.
- You must remember back then, churches were very different. They used to brew and sell ales etc. (remember the monks back up the road at the Coopers Arms!).
- This brewing and selling was a vital fundraiser for the church.
- So what would be better than after an ale or two at a Christening or fete etc., than a bit of entertainment, namely Morris Dancing?

- Now jump to the mid-1600s, and 'Morys Dancing' starts to appear the London's Lord Mayor Show/procession.
- From here on in, it becomes a staple requirement for folk festivals etc.

The Two Brewers

Address: 113 High St, Rochester ME1 1JS

Hours:

Monday	—	Closed
Tuesday	—	4:00 pm–11:00 pm
Wednesday	—	12:00 pm–11:00 pm
Thursday	—	12:00 pm–11:00 pm
Friday	—	12:00 pm–11:00 pm
Saturday	—	12:00 pm–12:00 am
Sunday	—	12:00 pm–7:00 pm

Established during the reign of Charles II (I really like this King, definitely one to hold a grudge).

- After the death of Crowell and England returned to being a monarchy, King Charles II was instated as King.
- In January 1661, the King, through parliament, ordered the digging up of Oliver Cromwell's body interred at Westminster Abbey, along with two of his mates.
- These three were then posthumously executed (hanged) at Tyburn at left there all day.
- Eventually, he had them cut down (nice of him!), but wait.
- Not wishing to come across as benevolent, he then proceeded to have their heads cut off.
- Erm, what could he do next? I know; let's put their heads on long poles outside Westminster Hall.
- Why Westminster Hall? I hear you ask; well, that is where his dad's trial was held and where he was sentenced to death.

- Maybe, just maybe, Charles II did hold a grudge after all. Cromwell's head was left there for at least 20 years.

Back to the pub, erected in 1683 by a guy called Thomas Grimmit, who had previously bought the timbers from Rochester Castle.

Thomas Preston was the first keeper of the Inn and was also known as the 'Brewster', a beer retailer and cooper of Rochester.

Preston examined the casks and the ale within and by marking them in grades with either one, two, or three crosses.

In 1775, extensive works were carried out to the Inn, internally and externally. The present facade was erected during this period. Alterations were also carried out in the mid-nineteenth century, it was during these alterations that an antique wooden peg tankard was found, said to have belonged to Thomas Preston.

Route: Nearly there! Turn right out of the pub and continue down the High Street till you reach the final watering hole on this crawl, The George Vaults, which will appear on your right.

On your way down, if you are eagle-eyed, you will see a plaque about head height on your right-hand side, midway between Deanery Gate and College Lane. It is entitled;

Abdication House

What in recent times was a bank, but back in history, this was not always the case.

In 1688 as the plaque recalls, "King James II of England and VII of Scotland stayed at this house as the guest of Sir Richard Head before embarking for France on 25 December 1688 when he finally left England."

A King Without a Crown: James VII and II's years in exile.

This short stay was due to him being replaced on the throne of England, by his daughter Queen Mary II and her King William III, following an uprising known as the Glorious Revolution (they then, both ruled as Joint Monarchs).

This removal was dressed up as an abdication by the Convention Parliament the following year, but in reality, it was either a coup or invasion, depending on where you sit.

William III (William of Orange) was also known for passing power from the monarchy to parliament.

As you continue down behind the shops, on your left is Rochester Cathedral.

So let's talk about it and start with:

Cathedral Green Men

- If you lift up your eyes up to the ceiling near the organ, you may glimpse the Cathedral's Green Men. (Remember them from earlier?)
- So it is assumed that in the 500s when missionaries arrived in Rochester from the Holy See of Rome, they used a bit of skulduggery by using the Green Man, which assisted in the conversion of the people as it provided a common thread for them to latch on to.

- New meanings were then attached to the images over time, such as 'new life', not just fertility etc. Indeed, this then allows the term resurrection to be slipped into common parlance.
- The ugly faces of the Green Men were also meant to protect the Cathedral from malicious spirits.

Just a snippet to ponder on your crawl, if you ever fancy undertaking a more serious crawl, i.e. pilgrimage;

The Pilgrims Way and Rochester

What is the pilgrims' way? It is a nearly 900-year-old trek from Winchester Cathedral (or Southwark Cathedral; shorter trip) to Canterbury Cathedral.

It started as a homage to Thomas a Becket, who was brutally murdered in Canterbury Cathedral by four assassins who had travelled over from Normandy especially.

Why? Well, it was all a big mistake! (if you believe that). All Henry II said was, "Will no one rid me of this troublesome priest," not butcher the vicar!

What had Thomas done for the King to be so miffed?

Well, he excommunicated some Bishops, whom he and the Pope thought had overstepped their mark after they performed the coronation of the young King while his father still lived. And that was Thomas's job, after all.

Soon after his murder, his burial site and relics became the place to visit on a 'pilgrimage'.

But, there is a spur to this pilgrimage that Pilgrims can do, and that is to slip over to Rochester Cathedral before going on to Canterbury.

Geoffrey Chaucer mentions this trek in his book Canterbury Tales.

It recounts the walk of a group of 31 Pilgrims (they take the shorter trip from Southwark, which is known as The Becket Way) who meet at the Tabard Inn, and while en-route each regale two stories each while on their walk, and the rest they say is history.

Ghost Alert

Before we get to the George Vaults, this part of the street has had a lot of spectres reported as being active over the years.

The houses/shops at 30/34 have unexpected visitors, so be on the lookout if you go shopping. So be aware that if you bump into a Roman Centurion walking along here, it may not be a fancy dress!

53 High Street: Dickens House

- Built in 1742. This seems to have been given its name as it is believed Charles Dickens stayed here on many occasions.
- A bust of his head is above the door, and a previous owner stated he saw Charles's ghost on many occasions.
- Smells are often used to indicate a spectral presence, especially if the smell is entirely out of context.
- That is the case here, where owners would regularly smell burning tobacco, yet no one smokes., Oh, wait Charles Dickens did, and he was a big fan of Syrian

Latakia: (Latakia is sited at the base of Mount Lebanon).

44 High Street Rochester (The Wonky House)

- Recently, a couple quickly decided to take over this more than 500-year-old wonky house.
- This house has been on High Street for over 500 years, but 400 years ago, it was a house of 'ill repute' which supported the 'oldest profession in the world'.
- After only a few days, unexplained goings-on started happening.
- Unbeknownst to this couple, it was local knowledge that this was a haunted spot.
- Anecdotally, the tale goes that a 'customer' became a little bit 'too handy' with one of the ladies.
- His ardour was cooled when he had his fingertips removed and thrown into the basement.
- So after moving in, Lynne (who had very nice kept nails and hands) noticed her fingers started to become engorged, and had little cuts appear.
- It got so bad she thought her fingers had become necrotic.
- Her hubby James also noticed a change in his finger's appearance.
- To accompany this phenomenon, this Lemur/Poltergeist also took to throwing jars, interrupting the electricity supply and blaring classical music from the radio.

- So, if you fancy popping in on your way to the pub, feel free, but take care, and count your fingers on your way out.

The George Vaults

Address: 35 High St, Rochester

Hours:

Monday	—	9:00 am–11:00 pm
Tuesday	—	9:00 am–11:00 pm
Wednesday	—	9:00 am–11:00 pm
Thursday	—	9:00 am–12:00 am
Friday	—	9:00 am–1:00 am
Saturday	—	9:00 am–2:00 am
Sunday	—	9:00 am–11:00 pm

This pub is a real historical find (looking at the front, 'never judge a book by its cover'). Rebuilt after a fire burnt down the pub (then called the George and Dragon) in 1768. It was built over an undercroft in 1335 and had associations with Bishop Gundulph. This fact is what makes the building historically interesting. If you get a chance, look at the arched ceilings and imagine what it would have looked like back in the 1400s.

Although no evidence can be seen today, it is believed by some that underground routes existed from here to the Castle and the Cathedral.

Ghost Alert: The Dark Monk

As you enter the vaults, you may find the air slightly chilly. This will have nothing to do with the heating. If you feel uneasy, it may just be you are being watched by the Black Monk, no one knows his back story, but you can imagine it may not be the thing that puts you at ease, staring into a faceless hooded habit.

Other spectres reported here include a friendly old gent, who appears, smiles, and then leaves again.

A previous landlord stated that the apparitions only started appearing after the tunnels had been closed and filled in.

And talking of landlords, it is rumoured that way back, a landlord committed suicide upstairs by hanging himself. Always a good starting point for a haunting, and yes, some witnesses have seen the shapes of people swinging from a rope.

We have now reached the end of this crawl, but as you leave Rochester, here are a quick couple of other interesting things you might want to see;

Back to Confusing Postman Pat

- Now we saw the slightly different one outside the Post Office, but down this end of the High Street, there are two others in place outside the Guildhall Museum.
- Both are believed to have been here since Victorian times.
- One green and one black.
- The colour Black is used to denote, 'not in working order' or 'not to be used'. This is to prevent confusion.
- Strange but true.
- England's very first post box installed was up north in Carlisle. Although no longer still there, there is a copy in its place to mark the spot.
- But the most prestigious historical ones are those made by John Penfold.

- These were easily recognisable by their hexagonal design and topped off with Acanthus buds.
- They mainly came in three sizes, but apparently, the aficionados' say there were 13 differing variants!
- In fact, there are only a dozen or so still in place, and Rochester has one!

Now in another second (remember the King's School earlier!), let me present to you:

England's Largest Pre-owned Bookshop; Well Almost!

- Baggins Book Bazaar (19 High Street)
- Rochester claims to have England's largest second-hand bookshop.
- There is almost like a Hogwarts maze of aisles, stairs, piled high books and discreet reading areas.
- Upon entry, you are greeted with a sign which asks you to leave bags at the cashier point while you wander to your heart's content around Rochester's answer to the Library of Congress.
- You may find this request a little off-putting and even accusatory (thinking you may be trying to slip a book into your bag!), but nothing could be further from the truth. As you stroll through these paper tunnels, you soon realise that the width of the paths and the gaps between the high piles of books gradually reduces. So not carrying a bag or a backpack makes complete sense.

So let's head off now the Stratford upon Avon.

Crawl 4
Stratford Upon Avon

Name Origin:

Stratford combines the old English street and ford, indicating a shallow part of a river. The 'street' referred to was a Roman road from Icknield Street in Alcester to Fosse Way. The ford bit in the name later became the location of Clopton Bridge.

In the late twelfth century, Richard I (Lion Heart) granted a charter, permitting the town to hold a weekly market. Thank God for that, so now we can explore that here;

- Records show that in 1251 Stratford was used for the first time to denote the area around the Holy Trinity Church and parts of the Old Town.
- A plethora of the town's earliest buildings are located throughout what is known locally as Stratford's Historic Spine.
- This followed a route from the town's centre to the Holy Trinity Church.
- Starting off where the 'Bard' was born on Henley Street, flowing into High Street (Harvard House) into

Church Street (Guild Buildings etc.) and ending up at the Holy Trinity Church, where William Shakespeare is buried.

Christmas Market (Various Streets)

This town's award-winning Victorian Christmas market spreads traditional festive cheer with plenty of shopping, food and street entertainment to enjoy. Over 300 stalls, many of which are being staffed by Victorian-clad traders. Selling a wide range of gifts and seasonal products, and although you may find this busy, as it is only on specific days, it is definitely worth visiting. I suggest you try and book somewhere to eat before you arrive as the pubs/restaurants get oversubscribed very quickly.

For location times and dates etc., visit the following website: *https://www.stratford.gov.uk/markets/stratford-upon-avon-victorian-christmas-market.cfm*

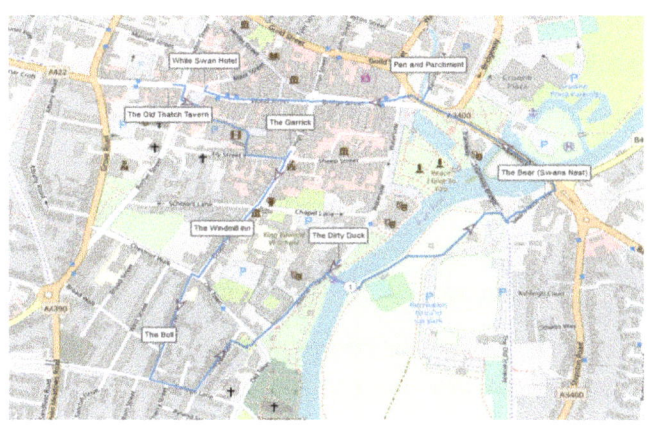

https://www.plotaroute.com/route/1962034

Pubs on this crawl:

The Old Thatch Tavern	—	(Greenhill St, Stratford-upon-Avon CV37 6LE)
The Garrick	—	25 High St, Stratford-upon-Avon CV37 6AU
Windmill	—	22 Church St Stratford-upon-Avon CV37 6HB
The Bull	—	9 Bull Street Stratford-upon-Avon CV37 6DT
The Dirty Duck	—	Waterside, Stratford-upon-Avon CV37 6BA
The Bear	—	Swans Nest, Stratford-upon-Avon CV37 7LS
The Pen and Parchment	—	Bridge Foot, Stratford-upon-Avon CV37 6YY
The White Swan Hotel	—	Rother St, Stratford-upon-Avon CV37 6NH

About 1.8 miles

The Old Thatch Tavern

Address: Greenhill St, Stratford-upon-Avon CV37 6LE
Phone: 01789 295216

Hours:

Sunday	12:00 pm–6:00 pm
Monday	12:00 pm–11:00 pm
Tuesday	12:00 pm–11:00 pm
Wednesday	12:00 pm–11:00 pm
Thursday	12:00 pm–11:00 pm
Friday	12:00 pm–11:00 pm
Saturday	12:00 pm–11:00 pm

This is a traditional old English Pub, which has had a licence since 1623 but has been on this site since 1470 as a pub. If you consider that in 1552, an Act was passed that meant innkeepers had to have a licence to run their pub. This one managed to get by for another 70 years without. Mind you, estimations are that in 1522 there were over 19000 pubs, alehouses and inns. Therefore it may have been difficult to get around and check them all.

What is also amazing, is that this is the only thatched building still in the town centre and is Grade II listed.

Ghost Alert

It was Christmas Day 1795 when Joseph (or James) Pinfield was being chased (allegedly) by some Irishmen. He ran into the tavern to seek refuge, but he may have been better off running into a church and shouting 'Sanctuary' because the mob found him in the pub, dragged him outside and eventually killed him. But Joseph appears to have liked the pub so much, he has decided to remain there and become the resident ghost!

Route: As you leave the pub, turn right into Rother Street. Continue along Rother Street until you reach Ely Street, which is the third turning on your left. Follow Ely Street till it joins Chapel Street, here you turn left into the High Street, and the Garrick is just down on your left.

As you go along Rother Street, you will see the Stratford PlayHouse on your right.

- Rother Street was established when laying out Stratford as a town in 1196.
- The word Rother comes from the Anglo-Saxon word meaning cow, and spookily it became home to the local cattle market.
- Some six centuries later, two fine buildings were erected on the site of nos. 14 and 15.
- This was then known according to records as Rother House.
- Ownership and usage changed over the forthcoming years, including a priest's house, chapel, library, convalescent home and even a maternity hospital circa 1940.
- When the NHS was formed in 1948, it was sold to the predecessor of the Stratford town trust for £30K.
- Eventually, we get to the latest incarnation as a playhouse, and it is a great venue to catch a production.

Christmassy Bit:

'Silent Night, Holy Night', some people's absolute epitome of the Christmas Carol, has its origins steeped in a story of poverty and despair.

- Legend has it (actually, it is a fact) that 'Stille Nacht' had its first airing on 24 December 1818.
- The venue: St Nicholas Church (near Salzburg, Austria).
- Let me now introduce you to Joseph Mohr, a young Catholic priest based at St Nicks (entirely appropriate, being Christmas and all!).
- But all was not well; Joseph was at a loss, the organ's bellows had been breached by very hungry mice, and there was zero opportunity for a repair before Christmas Eve Service later that day.
- Now, an electric light bulb moment, (could it have been divine intervention?), whether luck or whatever, we are all grateful for what happened next.
- In 1816, Joseph wrote a poem of six stanzas called Stille Nacht. Thinking quickly on his feet, he contacted a local organist and schoolmaster (Franz Gruber). He asked him if there was any chance he could add some music to his poem on the quick.
- So that night, Joseph and Franz sang 'Stille Nacht' for the very first time at Christmas Mass.
- Many translations later (and after some stanzas being omitted), we have today probably the world's favourite carol.

As it is only a short hop to the next pub, here is a timeline of some facts about Stratford upon Avon.

1557: A great year this, as this was the year that William Shakespeare's parents, John Shakespeare (Wool dealer and a future Mayor of Stratford) and Mary Arden (daughter of a yeoman farmer: [yeoman farmer meant he owned his own land]), got married here in Stratford.

1564: 7 years after their marriage, young William was born.

1582: Aged 18 William made an honest woman of Anne Hathaway.

1597: 15 years later, the couple buy a house on the corner of Chapel Land and Chapel Street.

1607: Now we have an American bit of history. This year at Harvard House, John Harvard was born. He became very famous in America, as it was he who gave his name to Harvard University.

The Garrick Pub

Address: 25 High St, Stratford-upon-Avon CV37 6AU
Phone: 01789 292186
Hours:

Thursday	—	11:00 am–11:00 pm
Friday	—	11:00 am–11:00 pm
Saturday	—	11:00 am–11:00 pm
Sunday	—	11:00 am–11:00 pm

Monday	—	11:00 am–11:00 pm
Tuesday	—	11:00 am–11:00 pm
Wednesday	—	11:00 am–11:00 pm

The Garrick Inn remains one of the oldest buildings in the town, dating back to the sixteenth century. A traditional half-timbered building and some parts of it, however, are rumoured to be even older and from the fourteenth century.

Previously called The Greyhound and the Reindeer but was renamed in honour of famed Shakespearean actor Sir David Garrick in 1795. David did much to popularise Shakespeare in his day and is credited as the architect of the Shakespearian tourism around the town.

Three storeys high, with the upper floor slightly projecting, epitomises typical Elizabethan townhouses.

The main building is likely to have been built at the end of the 1500s, although there are some parts that date back to the 1300s. But, since 1718, it has been in constant use as a pub.

On the darker side, a plague may have started within the original building in 1564. However, although an apprentice did die of the plague, it's unclear whether or not the phrase 'hic incepit pestis' ('here begins the plague') referred to the specific 'patient and place zero' or was added just as a general reference to the burial register.

Ghost Alert

It is a little non-specific as to who returns to haunt the Garrick, but staff and customers have reported, in the past, eerie feelings, apparitions etc. Now for a pub/building 700 years old, there must have been many a visitor who liked it so much, has refused to leave. Rumours include that Oliver Gunn may be one of the spooky visitors, or maybe the resident of the disturbed tomb (which was under the building), and has been released forever. Either way, aficionados of the spectral world agree something is going on here.

So when you are sipping your ale, and looking around in awe at this old pub, just make sure in the 'dimmed' light the people you think you see are indeed actually real people!

Route: Coming out of the Garrick turn right and go ahead down Chapel Street, continue into Church Street and the Windmill is down there on your left.

On your way, on the right-hand side, you will see Sheep Street:
Sheep Street

Christmas Quote:

"Christmas magic is silent. You don't hear it. You feel it. You know it. You remember it."

– Kevin Alan Milne
The Paper Bag Christmas

- Sheep Street has a history of its own, having buildings here from 1480, and some following the fire of 1595, many more were rebuilt.

- If you cannot guess from the name Sheep Street, this area was used as a marketplace for buying and selling sheep from the nearby Cotswolds.
- So let me draw your attention to Shrieves House. This is one house which has been here for years and is still lived in today.
- William Shakespeare is believed to use a former resident as the basis of one of his most loved characters Sir John Falstaff, who appears in three of his plays and is spoken of in a fourth.
- It is also a common belief that in 1651 Oliver Cromwell stayed here in and wrote a letter to Lord Wharton before the Battle of Worcester.

Shrieve's House

Touted as the oldest lived-in house in Stratford, the first building was on this site in 1196.

The Civil War

- In 1642, just before the Battle of Edgehill (which was the first major conflict of the Civil War), Dutch mercenaries (paid for by the Parliamentarians) bunked down where the museum now stands.
- During this time, a local lad called Lee was fascinated with foreign soldiers/mercenaries.
- At 14, seeing the arrival of this army, only proved to stoke his interest. He was often observed taking an avid interest in their drills/movements from a nearby pub.

- But unfortunately, as the date of the battle neared, tensions/suspicions were heightened, and a rumour that Lee was a Royalist spy sealed his fate.

- He was captured by the soldiers and quizzed using various torture techniques.

- Now you can imagine, just before a battle, fear must play a part in one's mind-set. Indeed this fear allegedly drove Lee's captors on in an effort to get information. However, it ended in a bad way for Lee.

- It is believed he was hanged from an upstairs beam (over the staircase) using a length of rope.

- His lifeless body was left for all to see until a senior officer cut him down.

- Disgusted with what his mercenaries had done, he quickly arranged for his body to be buried in a small shallow grave in the basement.

- Now jump forward 200 years or so. During some building works, they found a shallow grave which contained the mutilated remains of a teenage boy. Showing all the signs of being crushed into a too smaller grave.

Ghost Alert

- Shrieve's House (40 Sheep Street) is renowned for its spooky and unexplained ghostly happenings.

- As it appears to have been here for centuries, surviving the Civil War, the Plague and a major fire, it has a lot of stories to tell.

- The house is known locally for having at least three ghostly residents.
- One is an older woman who has been seen struggling to climb the stairs, carrying a candle in what appears to be a deformed hand.
- Another is a soldier who moves unexpectedly quietly about the house, bringing with him a spine-chilling aura/sensation.
- Let's now talk but the barn! This area has been known to leave people feeling claustrophobic and hemmed in. Many thought that an invisible hand was at the core of the sensation.

More Ghosts

Just down Chapel Road, on your left, you will see The Shakespeare Hotel. Now here, it is rumoured a few paranormal entities exist.

- One legend recounts the life of a local pickpocket, aptly named Lucy (an old nursery rhyme springs to mind, Lucy Lockett lost her pocket).
- Back then, in the 1300s, after a day swiping in the streets of Stratford, she would retire to her home. This happened to be on the site where the Shakespeare Hotel now stands.
- However, Lucy's lot was not a happy one (Gilbert and Sullivan). The story continues that her uncle not only sexually abused her but killed her and hid the body by building it into the walls (roughly where room 203 now stands).

- Now we all know how stories gain momentum, and these are amplified by staff being scared to enter on their own.
- There was another spooky sitting when an air hostess apparently awoke in room 203 to see a girl standing, just looking at her. As she got up from her bed, the apparition faded but left an impression on the young lady.

We should nearly be at the next watering hole by now, but here is a Christmas factoid.

Christmas Quote:

"The smells of Christmas are the smells of childhood."

– Richard Paul Evans

Windmill

Address: 22 Church Street, CV37 6HB
Hours:

Day		Hours
Monday	—	11.00 am–11.00 pm
Tuesday	—	11.00 am–11.00 pm
Wednesday	—	11.00 am–11.00 pm
Thursday	—	11.00 am–11.00 pm
Friday	—	11.00 am–Midnight
Saturday	—	11.00 am–Midnight
Sunday	—	11.00 am–11.00 pm

The Windmill has been here in Stratford since 1599 and claims to be the town's oldest pub. With its timber-framed walls and low ceilings, it exudes character and atmosphere and is definitely Olde-Worlde. It has a large bar area made from several smaller rooms, plus a lounge on the right, with an outdoor patio/smokers area at the rear.

Route: Turn left out of the pub and continue down Church Street. At the T junction with Chestnut Walk, turn right, then immediately turn left into Bull Street. Straight down Bull Street, and the Bull will be on your left.

So as we go on our route, let's talk Fire!

Fires in Stratford in 1594, 1595 and 1614

- Now I do like a fire, and these old houses presented many an opportunity for a conflagration to start.
- One such fire started in May 1594, destroying over 100 houses and barns in the town centre.
- As you can imagine, back then, many local industries used fire to prepare their wares or form their products.
- These workers included blacksmiths, bakers, cobblers, tallow chandlers, maltsters and brewers.
- On the day of the fire, records recall that the weather (high winds) whipped up the flames and spread quickly amongst the thatched roofs, which were very close to one another.
- You would have thought this fire may have prompted some 'fire prevention work', but no, a year later, in July, another fire occurred and spread rapidly through

another 20 houses. However, the fire was stopped or ran out of momentum before reaching the house, which was eventually bought by the Bard in New Place in 1597.

- Over the following few years, lawmakers put into place a by-law in 1612 which required businesses to have proper chimneys, and no thatched house could remain in the town centre.

- However, two years later, another major conflagration occurred, tearing through at least 54 houses plus additional outhouses and barns!

Christmas Quote:

"Christmas is doing a little extra something for someone else."

– Charles M. Schultz

Some Christmassy Bits

So let's talk Rudolf, that's right, the ninth and most special Reindeer of Santa's Crew.

Why Rudolf? Before we get to the whys and wherefores, you need a bit of the backstory.

- It was 1939, and the great depression in America was beginning to end. America had not yet been drawn into WWII.

- Montgomery Ward (retail and Catalogue Company) of Chicago decided their store should make its

Christmas Holiday book for children, as a promotional tool.

- So the manager approached one of their advertisement team and asked to come up with something a bit special.
- Robert May was known to have a laugh and was good at retelling tales etc.
- As with many 'perceived extroverts', when you scratch under the surface, there lies an introvert who lacks confidence.
- Robert was no different. He always felt like an outcast/underdog and felt he was not reaching his full potential.
- Even as a child, his favourite story was that of the Ugly Duckling.
- For Robert, though, this timing was also tricky, as unbeknownst to his boss etc. Robert was going through a bit of personal grief as his wife was dying.
- So, the story was developed. Initially, it did not impress the boss, but he persisted with the theme.
- Halfway through the project, Robert's wife died, and his boss offered to take the project off of him, but Robert insisted that he needed Rudolph.
- He chose the name Rudolph for alliteration reasons, it is believed.
- So if you think about it, Rudolph, being different and being made fun of by his peers, actually reflects the ethos behind the Ugly Duckling and promotes a message of the belief that everyone has something to offer. It's just finding your niche.

- Apparently, Robert did not own the rights to Rudolph, but Montgomery Ward (in fact, they never really saw it as anything but a promotional tool), so after the war, they signed over the rights to Robert.
- Now comes the tsunami. In 1949, Robert's brother-in-law wrote the song, which was picked up by Gene Autry (singing cowboy) and sold over 25 million copies. This was the genesis of the TV film show.
- Rudolph the Red Nose Reindeer original film was voted by 83% of those that took part, in the Hollywood Reporter/Morning Consult poll, as the 'most beloved holiday film'.
- The rest as they say (and even in the song) is Christmas history.

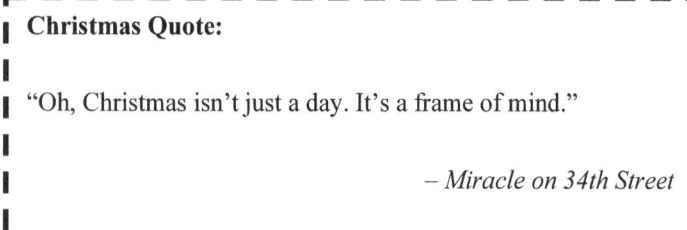

Christmas Quote:

"Oh, Christmas isn't just a day. It's a frame of mind."

– Miracle on 34th Street

The Bull

(Formerly, West End)
Address: 9 Bull Street, CV37 6DT
Phone: (01789) 471931
Hours:

Monday	—	Noon–11.00 pm or later
Tuesday	—	Noon–11.00 pm or later
Wednesday	—	Noon–11.00 pm or later
Thursday	—	Noon–11.00 pm or later
Friday	—	Noon–11.00 pm or later
Saturday	—	Noon–11.00 pm or later
Sunday	—	Noon–11.00 pm or later

About the Pub

This old pub tucked away in the narrow backstreets of Stratford's Old Town has been known as the West End at various times over the years but was taken over in 2020 by the local Melt Pub Company and has reverted to its old name once again. Live music and other events. Garden/large heated patio.

Route: Turn left out of the pub and continue to the T junction with College Lane. Turn left into College Lane and take the second left into College Street. Continue past the Memorial Gardens turning right into Old Town. Take first left into Sothern Lane and continue along here till you reach the next pub, the Dirty Duck, on your left-hand side.

So what do most people associate with Stratford upon Avon? (Apart from Christmas markets) I guess it's William Shakespeare.

When Shakespeare was born on 23 April 1564 (Bapt. 26 April 1564) and died on 23 April 1616. The exact birthdate is not known but is assumed to be three days before the baptism, as that appeared to be the 'norm' back then. So he was born and died on the day we Celebrate St George's Day!

Back then, life expectancy was about 30 years, and less than half survived passed their middle teenage years.

William was born to John Shakespeare and Mary Arden. He was one of eight children, and he seemed to be born lucky, as within a few weeks of being born, the 'Black Death' (aka The Plague) ran roughshod through the town.

If your house was unfortunate to have people who had contracted the Black Death, a Plague Cross was sometimes used to highlight the fact. This was borne out by Samuel Pepys, who in 1665 wrote:

"I did in Drury-lane see two or three houses marked with a red cross upon the doors, and 'Lord have mercy upon us' writ there—which was a sad sight to me, being the first of that kind that to my remembrance I ever saw."

Diary entries from June 1665: The Diary of Samuel Pepys. In fact, in the last half of 1664, nearly a third of the town caught the plague.

Some factoids about the Bard!

- Shakespeare's family home was on the corner of Chapel Street and Chapel Lane, now known as New Place.
- The Bards tomb lies not far away in Holy Trinity Church.
- But as this was a time of superstition etc., a curse was added to the tomb to warn off people from stealing or interfering with his remains. It warns "Good friend, for Jesus' sake forbear, to dig the dust enclosed here. Blessed be the man that spares these stones, and cursed be he that moves my bones." (Facts About the End of Shakespeare's Life—ThoughtCo.) Although some legends have it that his Skull may have already been removed by some scoundrels.
- William had seven brothers and sisters: Joan; Margaret; Gilbert; Joan (No.2); Anne; Richard, plus Edmund, all born between 1558 and 1580.
- When William got married at 18, his wife (who was a tad older than him!) was already pregnant.
- They had a son and gave him the name of Hamnet, but the lad died at only 11 years old.

Shakespeare had his own family coat of arms, seen below has the Latin inscription 'Non Sans Droict', which means 'Not without Right'.

NON SANS DROICT

56 Fun William Shakespeare Facts About His Life & Works.

- Not a lot of people actually realise Shakespeare was not only an Elizabethan playwright, but as most of his better-loved plays were actually during the reign of James I, he was more of a Jacobean writer.
- It can get confusing here, but King James VI of Scotland came to be King James I of England when Lizzie died in 1603.

- So overnight in England, we went from being Elizabethan to being Jacobean. The Jacobean Era covers the time when James I ruled England.
- Jacobian translates as 'of James', and later on, the followers of James II were Jacobites.
- But it doesn't end there, William survived another monarch and died when Charles I was on the throne, so in fact, he was Elizabethan, Jacobean and Carolean!
- Everyone would expect that his whole family were loyal to the throne. But that was not the case, as it appears a relative of his, Mr William Arden, was arrested and executed for plotting to kill Queen Elizabeth I.
- Arden was related to William Shakespeare's mum.
- How involved he was has some doubt attached, as it was his son-in-law (not Will's dad), John Sommerville, who hatched the plot but was discovered before it ever really got going!
- Elizabeth (always one of leniency) set about retribution. William Arden was executed on 20 December 1583, the day after John Sommerville was found dead in his cell at Newgate. Officially declared suicide, but some are convinced he was killed.
- John Somerville had devised the plan to end the reign of the Virgin Queen by killing her. However, this was foiled before he ever got near to carrying out his fantasies.

Back to The Bard and another fire!

This time not in Stratford but the Globe in London.

- Although Shakespeare's Globe came to an abrupt end, it literally went out with a bang!
- On 29 June 1613, during the performance of Henry VIII, some small cannons were fired. It should be noted here that other theatres at the time were vying for audiences, and theatres did everything to try and make their productions bigger and better, so using a cannon was a significant special effect. However, they didn't use cannonballs. They did, however, use gunpowder and wadding.
- Unfortunately, a stray burning ember landed on the thatch roof, and bang! Major special effect.
- In fact, it took under two hours for the whole theatre to be raised to the ground.
- Luckily, nobody was severely injured. Although it was reported that a man's trousers/breeches caught alight, a nearby Good Samaritan doused the flames out with his Beer! (See Beer is Your Friend, and It Saves Lives.)

Well, enough of William Shakespeare (I am sure other bits will pop up now and again). You have enough to wow your friends, so let us get on with the serious business of drinking, I mean 'Crawling'.

On your way, you will come across the Clopton Chapel (this guy is quite popular around these parts!)

The Clopton Chapel is inside the Holy Trinity Church.

Who was this Clopton?

- Sir Hugh Clopton, who eventually became Lord Mayor of London in 1491, was fully committed to Stratford.
- He completely rebuilt the Chapel of the Guild of the Holy Cross.
- Also paid for the stone bridge over the Avon, which carries his name.
- This is Bard's final resting place.

Ghost Alert

Clopton House is a majestic manor that stands in Stratford-upon-Avon. It was home, at a time, to the Cloptons, who were a powerful family in Stratford.

The house has seen its fair share of tragedies, with a good example being the suicide of Margaret Clopton, who flung herself into the well of the house after getting news of her betrothed absconding with another woman.

But it's most notably believed to be haunted by the ghost of Charlotte Clopton. A girl who died after she was accidentally buried alive. She was only discovered sometime later when another family member died. Since then, it is said that you can still hear her screams as she tries to escape the vault that she was wrongfully buried in.

Let's Have Some Christmas Bits to Be Going on With

- Christmas songs can be quite lucrative! Mariah Carey makes about £375,000 yearly from her hit *All I Want for Christmas*. Shane and The Pogues make

- approximately £400,000 for their Fairy-tale of New York.
- However, the top of the pops is Slade. It is believed they earn in excess of 500K per annum from their standard Merry Christmas Everybody.
- The tradition of putting up Christmas trees dates back to the sixteenth century.
- Where did it all start? Popular belief is that the tradition that can be traced back to Germany. However, Tallinn (Estonia), and Riga (Latvia) may have something to say about that!
- Latvians have a legend that decorating Christmas trees began in Riga 500 years ago.
- In 1510, the Brotherhood of the Blackheads (a well-known local guild in Riga). Now, these Blackheads (a great read and worth some additional research if you are interested) were keen participants in community life.
- Historically, around the Winter Solstice in 1510, the brotherhood went into the local forest to locate and cut down the biggest fir tree they could find. The idea was to set the tree on fire on the banks of the Daugava River, reinforcing the old tradition of the burning log.
- However, the tree they found was humongous, and once it was back on the banks of the river, they felt it was too big to set fire to it as it may endanger some localised buildings.
- Worried about this, they returned to their offices to discuss what to do next.

- Meanwhile, local children saw the tree unattended and dressed it with whatever they could find.
- Children used the yarn from their gloves and added fruit and berries etc.
- As night approached, the kids bid the tree farewell and went home to tell their parents what they had been up to.
- Meanwhile, the Blackheads meeting ended with a resolution being found.
- Returning to the tree, the blackheads discovered what had been going on. They were taken aback, so much so, that they declared this to be The Christmas tree, and the rest, they say is history (or is it?)
- Now for Tallinn in Estonia. Their legend tells that in 1441 the Blackheads (again) brought a tree to Tallinn's Town Hall Square. This is some 60 years earlier than Riga. I suppose the only ones who really know are the Blackheads.
- Queen's Bohemian Rhapsody holds the record for getting to the U.K. Christmas Singles Chart Top Spot twice, in 1975 and 1991.
- Now some will say the above fact is incorrect as 'Do They Know It's Christmas' hit the top spot three times (1984, 1989 & 2004). However, the groups that sang it differed, so it doesn't count.
- Personally, I am a bit partial to some Elvis songs, especially his Christmas album. However, this view was not held by Mr Irving Berlin, who detested his version of White Christmas so much he tried to get radio stations to boycott it!

- There are more than 630 different species of Christmas trees, including the balsam fir, Douglas fir, and Fraser fir, to name but a few.

- Now when you watch the video, you would never guess that David Bowie hated the selection of 'The Little Drummer Boy' as the duet he was to perform with Bing Crosby. But a quick rejigging of the tunes by Kohan, Grossman and Fraser saved the day.

- We can all name Santa's reindeer, but did you know that two of the crew (originally) had different names?

- In Clement Moore's 1823 poem, A Visit from Saint Nicholas (T'was the Night Before Christmas), Donner and Blitzen were originally named Dunder and Blixem! The way the Dutch refer to 'thunder and lightning'.

- Mel Tormé's Christmas standard 'The Christmas Song', immortalised by Nat King Cole (aka 'Chestnuts Roasting on an Open Fire') was actually penned during the 1944 heatwave.

- We all love children at Christmas (and throughout the year!), But did you know that Brenda Lee's Christmas hit *Rockin' around the Christmas tree* was recorded when she was a child 13?

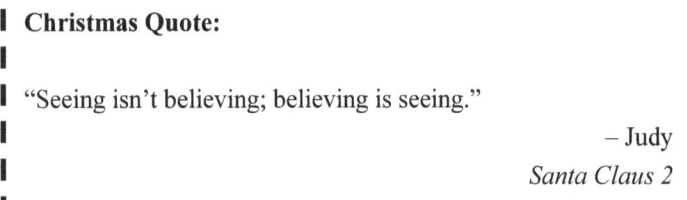

Christmas Quote:

"Seeing isn't believing; believing is seeing."

– Judy
Santa Claus 2

The Dirty Duck Pub

Address: Waterside, CV37 6BA
Phone: 01789 297312
Hours:

Monday	—	12:00 pm–11:00 pm
Tuesday	—	12:00 pm–11:00 pm
Wednesday	—	12:00 pm–11:00 pm
Thursday	—	12:00 pm–11:00 pm
Friday	—	12:00 pm–11:00 pm
Saturday	—	11:00 am–11:00 pm
Sunday	—	11:00 am–11:00 pm

Sitting on the waterside of the river Avon here in Stratford. Known also as the pub with two names, previously known as The Black Swan but latterly (during WWII), it became fondly known as the Dirty Duck by the soldiers from the USA who were camped nearby. However, other legends say it was given the name by locals, but I like the war version and could well imagine that happening. That said, it remains the only public house in England with two names on the licence.

Route: Turn right out of pub, and just down on your left (if you are lucky), you will find the chain ferry, which will take you across the river for about 20p. Once across, keep the river on your left and continue till you reach Swans Nest lane. The pub inside the hotel is just down here on your left.

As we go along, here are some more titbits about Will!

According to the Literary Encyclopaedia, Shakespeare is Number 2 among the most quoted English writers. Only kept off the top spot by the writers of the Bible.

Here is a list of a few phrases that you are well aware of but maybe not know originated from the pen of William Shakespeare.

1. "I must be cruel, only to be kind"—Hamlet
2. "One fell swoop"—Macbeth
3. "Too much of a good thing"—As You Like It
4. "Break the ice"—The Taming of the Shrew
5. "With bated breath"—Merchant of Venice
6. "A wild goose chase"— Romeo and Juliet
7. "The be-all and the end-all"—Macbeth
8. "Dead as a doornail"—Henry VI, Part II
9. "Not slept one wink"—Cymbeline
10. "I have been in such a pickle"—The Tempest
11. "Such stuff as dreams are made on"—The Tempest
12. "Eaten out of house and home"—Henry IV, Part II
13. "Kill with kindness"—The Taming of the Shrew
14. "Laugh yourself into stitches"—Twelfth Night
15. "It's Greek to me"—Julius Caesar
16. "Give the devil his due"—Henry IV, Part I
17. "Knock knock! Who's there?"— Macbeth
18. "All that glitters is not gold"—The Merchant of Venice
19. "I will wear my heart upon my sleeve"—Othello
20. "Brave new world"—The Tempest
21. "Wild-goose chase"—Romeo and Juliet

22. "Eaten me out of house and home"—Henry IV, Part II

23. "Jealousy is the green-eyed monster"—Othello

24. "Milk of human kindness"—Macbeth

25. "Though this be madness, yet there is method in it"—Hamlet

The complete works of William Shakespeare are available at http://shakespeare.mit.edu/.

Christmas Quote: (Those who know, know!)

"For many, Christmas is also a time for coming together. But for others, service will come first."

– Queen Elizabeth II

As we walk back and across the River Avon, here are some interesting bits about Shakespeare's Avon, or Warwickshire's Avon.

There are nine rivers which go by the name of Avon situated within Great Britain.

Stemming from an ancient language, the River Avon actually means River River

- So there are five Avon's in England, three in Scotland and one in Wales.
- However, the one in Wales is spelt as Afon Afan but still means River River.

ENGLAND Shakespeare's Avon

- This is the longest Avon in the British Isles, running a lengthy 85 miles through Leicestershire, Northamptonshire, Worcestershire, Warwickshire and Gloucestershire.
- Its source starts in Naseby, Northamptonshire and runs until it merges with the River Severn at Tewkesbury in Gloucestershire.

Bristol Avon

- AKA the Lower Avon, which is only 15 miles shorter than its counterpart in Warwickshire, travels seventy miles from its source at Acron Turville near Chipping Sodbury until it merges with the River Severn at its estuary at Avonmouth.
- As the crow flies, the distance from source to mouth is only 19 miles.
- However, due to its meandering, it reaches 75 miles and passes through Bath.
- The river has many branches and also merges with the River Kennet at Bath, which combines to form part of the 79-mile-long Kennet and Avon Canal.

Hampshire Avon or Salisbury Avon

- This is sixty miles long river and is different as it rises at two places, Devizes and Pewsey (both in Wiltshire) and joins together at Upavon.

- The river then meanders through the southern counties emptying into the English Channel in Dorset.
- A unique selling point about this River Avon is that it has the most species of fish in any British river.

Avon Water

- This nine-mile-long river rises in the New Forest and finally exits in the Solent.
- This is the shortest of the three major rivers in the New Forest.
- The other two are the River Lymington (14 miles) and Beaulieu Water (12 miles).

Dartmoor Avon

- Holds the prestigious prize as England's shortest River Avon, a mere seven miles long.
- Rising in Dartmoor National Park near Ryder's Hill, famous for its Standing Stones.
- After its relatively short journey, it empties out into the English Channel.
- It may be small (short), but it has a great reputation for the levels of Trout and Salmon which use this river.

SCOTLAND Cragganmui Avon

- So the granddaddy of the Avons is the one which rises at Cragganmui and runs for a total of 40 miles.
- However, after 10 miles, it enters Loch Avon, which is high up on the Cairngorm Plateau in the Cairngorms National Park.
- This River Avon is the longest branch of the River Spey and is held in high esteem by anglers who feel this is Scotland's best river for salmon.

Avon Water (Scotland)

- The 24 miles of this River Avon makes it the second longest in Scotland.
- Rising near Irvine in Ayrshire and winds a North Easterly path before joining into the River Clyde.

Darvel Avon

- This is Scotland's shortest Avon which rises at Darvel in Lanarkshire and flows for about 11 miles before emptying into the Firth of Forth.
- This river is unique as it is spanned by the Avon Aqueduct (which is 810 feet long).
- This aqueduct is the largest in Scotland and the second largest in Britain behind one in Wales.

WALES Afon Afan

- This rises in Cymer village and flows for 14 miles through a 40-acre Country park (Afan Argoed Forest Country Park) in the Vale of Glamorgan.
- It eventually empties out in Port Talbot into the Bristol Channel.
- Along its short route, you will discover two structures which are bound to impress.

One: Known as Red Bridge, and one is Y Bont Fawr (big bridge), fine example of eighteenth-century building and engineering.

So as we meander through the park towards the next pub, here are so more Christmassy Bits

Christmas tree facts:

1. Approximately 60 million Christmas Trees are grown and harvested each year throughout Europe.
2. In the UK alone, we use nearly 8 million real natural Christmas trees.
3. But as more and more people move to 'faux' trees, the number is slowly reducing, but still, the ratio of 3:1 is in the real tree's favour.
4. Did you know that Christmas trees may be good for you? In fact, lots of bits of the tree can be consumed, and if you are looking for a good source of Vitamin C, then look no further than the needles!
5. Some trees may be younger, but the average Christmas tree is grown for about 15 years.

6. To ensure supply is not disrupted, growers will plant up to three trees to replace each one which is harvested.

Right let's pop in for a bear, sorry I meant Beer!

Christmas Quote:

"Christmas is built upon a beautiful and intentional paradox; that the birth of the homeless should be celebrated in every home."

– G.K. Chesterton

The Bear

Located in: Swan's Nest Hotel
Address: Swans Nest, CV37 7LS
Phone: 01789 265540
Hours:

Monday	—	12:00 pm–11:00 pm
Tuesday	—	12:00 pm–11:00 pm
Wednesday	—	12:00 pm–11:00 pm
Thursday	—	12:00 pm–11:00 pm
Friday	—	12:00 pm–11:00 pm
Saturday	—	12:00 pm–11:00 pm
Sunday	—	12:00 am–10:30 pm

The Bear commands a unique riverside location with views over the Avon which will inspire.

Although it first opened in 1662, it has continued growing and remains relevant today.

Route: Turn left out of the pub and then cross Clopton Bridge (known as Bridge Foot). Continue straight until you reach the Pen and Parchment on your right. Junction of Bridge Foot and Bancroft Place.

Clopton Bridge

- This is a Late Medieval stone arched bridge with a total of 14 arches.
- Sited at the place where the Saxons forded the River (Strat Ford; Street Crossing).
- Built at the end of the 1400s and was paid for by Hugh Clopton of Clopton House.
- He later became Lord Mayor of London.
- This replaced the timber bridge there and was recorded some 200 years earlier.
- Restored in 1588 after significant flooding.
- More restoration occurred in around 1642 after parts of the bridge were demolished to halt the advancing army of Oliver Cromwell.

Sir Hugh and the Gunpowder Plot!

- A distant connection existed between the descendants of Sir Hugh and the attempted blowing up of Parliament in 1605.

- So some 115 years after Sir Hugh passed away, a gentleman known as Ambrose Rookwood, who was a second cousin in Sir Hugh's family, was put on trial & found guilty of being part of the plot and hanged until he nearly died then drawn and quartered on the last day of January 1606.
- Before his execution, while being imprisoned in the Tower of London, he scratched his name onto some stones in the Martin Tower.

As you now cross the river on your left, you will see Bancroft Gardens and the Royal Shakespeare Theatre nestling in the distance.

This is a great little area to explore, and within its grassed areas lies a Human Sundial dedicated to the Warwickshire Fire Service.

There are many more statues (including the Hermaphroditus and a swan fountain).

We will talk about the Gower memorial in a bit.

- The Bancroft was originally where the townsfolk grazed their animal.
- The Country Artists Fountain (this is the one with two Swans looking like they are engaged in some sort of courtship ritual) was created to celebrate it being 800 years since Richard the Lionheart (King Richard I) gave Stratford the Charter for Market Rights (1196).
- This is the work of Christine Lee and is constructed of brass and stainless steel.

- HRH Queen Elizabeth II removed the cover to show it to the world in 1996.

So to the Gower Memorial

- This is a Statue of The Bard seated on the edge of a chair holding a quill.
- At the base of his sculpture are four bronze wreaths, and each corner of the base is adorned with some ornate masks and some other bits, which I will describe as we go.
- What is different is that surrounding Shakespeare are four other small statues depicting characters and engravings from four of his plays.
  o Statue 1: History: Prince Henry, son of Henry IV [Mask has English Roses and French Lilies].
  o Statue 2: Comedy: Falstaff, a loveable rogue from three of the Bard's plays including, Henry V and Merry Wives of Windsor. (Mask has Hops and Roses).
  o Statue 3: Tragedy: Lady Macbeth. (Mask has Poppies and Peonies)
  o Statue 4: Philosophy: Hamlet: (Mask has Cypress and Ivy.)

On the statues, there are some inscriptions:
Hamlet
Good night sweet prince
and flights of angels

sing thee to thy rest
Macbeth
Life's but a walking shadow
a poor player
that struts and frets
his hour upon the stage
and then is heard no more
Henry IV
I am not only witty in
myself, but the cause
that wit is in other men
Henry V
Consideration like
an angel came
and whipt the offending
Adam out of him

So as is the way, and it is about Christmas as well, before we reach our next watering hole;

Christmas Bits!

Some fun facts about Christmas movies.

1. Yahoo held a poll where users voted the most popular Christmas movie of all time was Home Alone. Coming in second was The Muppet Christmas Carol, and my favourite, It's a Wonderful Life, took the bronze medal.

2. Hard to believe, but the FBI (in 1947) had some concerns about It's A Wonderful Life. A 'stupid' analyst thought the film was a blatant attempt to show

bankers as disreputable people, which was perceived at the time as a tool used by communists.

3. In It's A Wonderful Life, have you ever wondered why George Bailey looks hot and bothered when he is with Clarence on the bridge? That day, despite the artificial snow, it was actually 90 degrees.

4. So to the gold medal movie in this poll. Do you remember seeing the aesthetically challenged photo of Buzz's girlfriend? Well, it was actually the art director's son. They did not want to use a girl and set her up for distasteful remarks.

5. Remember the baddie, Marv? He had an irrational fear of spiders and insisted the 'spider on his face' was done in one take. Note also that the scream was added later to not scare the spider.

6. The other baddie Joe Pesci really was a baddie! And maybe he thought he was still filming *Goodfellas* because he would forget and drop 'fucks' all over the place during his outburst.

7. The film *Elf* may have been a different movie had the other actor (Jim Carrey) offered the lead accepted the role.

8. Now another bit of movie trivia, in Polar Express, when Tom Hanks (as the conductor) says '11344 Edbrooke', it's not a fake address but the childhood home of Robert Zemeckis in Chicago.

9. Another favourite of mine is White Christmas. But did you know that Rosemary Clooney, who plays Betty, is actually George Clooney's aunt?

The Pen and Parchment

Address: Bridge Foot, Stratford-upon-Avon CV37 6YY

Phone: 01789 297697

Open seven days a week 7:00 am till 11:00 pm

The Pen & Parchment is situated by the Avon and retains many original features, providing a tremendous atmospheric feel. They have included traditional elements, such as the original beams, and added contemporary touches, such as lovely leather sofas and bucket seats, providing comfort.

This has been a well-known landmark since the days of William Shakespeare. A bit of pub trivia and that's the wisteria winding at the front is 150 years old. One nice little fact is that you'll find inside, a wooden pillar taken from a ship, which was captained by Nelson.

Route: Come out of the pub, cross Bridge Foot into Bridge Street, and turn right (not back over the bridge!) Continue straight, and the road becomes Wood Street. Again continue here until you come to the next pub, The White Swan, on your right-hand side.

There is little to see on the direct route to the next pub, but going up Henley Street on your right side may be worth it if you want to detour. You can carry on till the end of Henley Street, then turn left down Winsor Street, when you reach our starting point turn left and the next pub is on your left.

So, Henley Street, you will find the Bard's birthplace here. Go here to learn more about him and his family, but here are some bits to keep you going.

- John Shakespeare—William's dad—lived and worked in the house with Mary Arden.
- John became Mayor in 1568, which was considered a high-status position. Some believe this enabled William to be accepted into the local Grammar School.
- William married Anne when he was 18, and they lived with his mum and dad. Indeed Anne gave birth to their three children, Susanna and the twins Judith and Hamnet.
- Will's dad died in 1601, and William being the eldest son, William inherited the house.
- William divided the property and leased the adjoining cottage to Joan Hart (his sister).
- What was left eventually became the Maidenhead after Will had leased that bit out too.
- After his death, the Bard bequeathed the home to his eldest girl Suzanna.
- She left it to her daughter Elizabeth, who, despite being married twice, produced no heirs.
- So after she passed, the house was left to a descendant of Will's sister Joan Hart.
- Some more passing and selling of the property occurred over the next few years till, in 1847, the house was bought at auction for three thousand pounds by the Shakespeare Birthplace Trust.

Ghost Alert! 21 Henley Street

- Just along the road from where William was born is 21 Henley Street. This is the last remaining part of an old inn which stood here originally, The White Lion Inn.
- This Tudor building was constructed in the sixteenth century, but there has been something here since the twelfth century.
- Back in the day, the White Lion was famous for being Europe's biggest coaching Inn.
- It's no real surprise that due to its age, size and position, it has a darker side.
- Indeed it is rumoured that the venue was frequented by some secret societies, and the Royals have also been linked to it.
- Maybe even Guy Fawkes had a drink here, who knows?
- Staying with the darker side of life, in the eighteenth century a series of brutal assaults took place on some of the ladies of the night who plied their wares in the bedrooms upstairs.

- Hardly surprising, then, that there still remain some malevolent auras around the place.
- At the time of the English Civil War, the Inn was taken over by soldiers loyal to parliament, and the spectres are still heard now and again as they whisper their plans on the upper levels.

Okay at the end of Henley Street is a statue called The Jester, this is worth a look either before our last stop.

Now Jesters were prevalent in days of yore indeed; Shakespeare used them in some of his plays. So here is a brief overview of Jesters and what other ones the Bard used.

- The Fool (Jester) in Elizabethan times and within Bard's dramas is a person whose sole purpose of his employment was to entertain.
- That could be in a palace court, for a king or queen or for anyone who needed entertainment.
- But contrary to our preconceived ideas, in Bard's time, the Fool is the cleverest man in the drama.
- Jesters should be distinct from the Clown.
- The Jester (Fool) plays two significant roles in Shakespeare's plays.
- He can say what he likes to the people in power, as he is a jester, whereas anyone who dares tells the king he has no clothes on would fear their lives (that's not Shakespeare, but you get the point).
- He can also assist the audience by providing commentary on the plot and characters.

So we know that Jesters (Fools) appear in four of Shakespeare's plays. We have already met Touchstone (more of him in a bit), but here are the others;

Feste (Twelfth Night)

- Probably one of the most famous/recognised of the Bard's fools.
- Although his role is to sing and dance etc., he is also an essential cog in the household of Olivia.
- His freedom to roam comes from the respect Olivia has for him.
- He can go where his mood takes him and even does some extracurricular work by entertaining other patrons.
- His character is brilliant with a 'particular bent' for languages.
- Differing from others within the Bard's Folio's Feste also has a dramatic role to play.

The Fool (King Lear)

- The fool without a name! In King Lear, he is called 'fool'.
- King Lear's ever-present companion sticks with him through thick and thin, even to the point where he is executed just before Lear's demise.
- It is almost like the Fool is Lear's inner Jiminy Cricket! He is the wise one, while Lear is the Fool.

- The Fool takes every opportunity to point out the King's follies.
- Finally, when Lear becomes self-aware that he is not as wise as he first thought, the need for the Fool is removed, so he dies.

Trinculo (The Tempest)

- Trinculo is more like the stereotypical fool, without wisdom and pretty stupid, almost like a clown.
- Trinculo is Alonso's court jester.
- They become shipwrecked, and the Fool teams up with Stephano (a drunk royal butler!) and Caliban (Prospero's servant).
- The threes incompetence in trying to hatch a plot to overcome Prospero allowed Shakespeare to write some of his funniest work.

Touchstone (As You Like It)

- This jester is Duke Frederick's provider of court entertainment.
- He, however, is no 'fool' but an astute commentator and observer of human behaviour.
- He also pulls no punches and consistently quips malevolently about other characters.
- His command of language (s) means he is very good at turning and twisting any argument.
- In fact, it is he, who talks about the wisdom of being foolish.

Now for Some More Christmassy Bits

Christmas facts: World Records

- The longest-ever Christmas stocking was 32.56 m long and 14.97 m wide.
- A preschool in Buckinghamshire holds the record for the biggest Christmas cracker. In 2001, it constructed one 63.1 m (207 ft) long and 4 m (13 ft) in diameter.
- The most Christmas cards sent by a single person in a year were 62,824.
- Do you wanna build a snowman? It would need to be bigger than 113 feet tall to be the biggest.
- In 2012, Reddit organised the biggest-ever Secret Santa with over 30,000 participants.
- We must return to 2001 to find the most expensive Christmas card. This was sold at auction for twenty thousand pounds.
- Staying with the money theme. It is estimated the most value attached to a Christmas tree fully decorated was just under £7 million. Can you guess where this was? Dubai in 2010.
- Carrying on with the Christmas tree theme, the town of Malmedy in Belgium holds the record for the most lights lit simultaneously on a Xmas tree, the number being 194,672! Spookily this was also done in 2010.
- Personally, this is a lot earlier for me, but according to Jarlesberg Cheeses' research (why the hell would they want to carry out this research!), the stress of getting the Christmas dinner ready, cooking, etc. pushes British people to have their first tipple 12

minutes before midday. They obviously never spoke to anyone who has 'Bucks Fizz' for their Christmas breakfast at 6:00 am, after being up for several hours with the kids opening their presents.

So onward to our last watering hole!

Christmas Quote:

"Christmas is the season of joy, of holiday greetings, exchanged, of gift-giving, and of families united."

– Norman Vincent Peale

The White Swan Hotel

Address: Rother St, Stratford-upon-Avon CV37 6NH
Phone: 01789 297022
Hours:

Monday	—	07:00 am–11:00 pm
Tuesday	—	07:00 am–11:00 pm
Wednesday	—	07:00 am–11:00 pm
Thursday	—	07:00 am–11:00 pm
Friday	—	07:00 am–11:00 pm
Saturday	—	12:00 pm–23:00 pm
Sunday	—	12:00 pm–22:30 pm

This lovely old pub/hotel dates back to 1450 and was used as a pub way back in 1560. Back then, it was known as the King's House (or Hall).

The family had some links to the bard, in that the owner's granddaughter's husband (I know it's tenuous!) Richard Tyler became one of Will's childhood mates.

It's not a hard stretch to think that the bard more than likely had a beer or two in here.

If you look around while sipping your beer, you may notice a wall painting. This fresco was probably commissioned by William Perrot and tells the story of Tobias and the Angel, where Tobias is sent to collect monies owed to his father, who was blind.

The big question, though, is this: Did Shakespeare sit and drink his mead while looking at that very painting?

So as you digest the atmosphere and history while supping, ponder one this, did Will and Anne do the same thing as you are doing at this very moment some 500 years ago?

So we have reached the end of this crawl, which like most of the crawls in this book, are slightly different to each other.

Because of Stratford's relationship with William Shakespeare, who influenced English Life, Culture and Language, it is hardly surprising that he features prominently in this crawl, and I make no apologies for this.

So as before we fly off to the next Christmas market, here are some last things about the Bard and Elizabethan England.

Christmas Quote:

"Christmas is a together sort of holiday."

– Pooh
Winnie the Pooh

Shakespeare and Christmas:

In all of his works, William only mentions Christmas three times! (Probably due to the fact that at the time Easter was the biggest Christian Celebration).

1. At Christmas, I no more desire a rose Than wish a snow in May's new-fangled mirth: *Love's Labour's Lost*, Act 1 Scene 1.
2. I see the trick won't: here was a consent, Knowing aforehand of our merriment, To dash it like a Christmas comedy: *Love's Labour's Lost*, Act 5 Scene 2.
3. Sly: Marry, I will; let them play it. Is not a commonly a Christmas gambol or a tumbling trick? Page: No, my good lord, it is more pleasing stuff: *'The Taming of the Shrew'*, Induction Scene 2.

Taming of the Shrew | Introduction 2.In fact, even his play *Twelfth Night* is not about Christmas at all although one might have thought it was by its title!

Christmas Quote:

"A Yule log. It's a wonderful tradition. One log is chosen, and everyone in the house touches it and makes a Christmas wish."

– Belle
Beauty and the Beast: The Enchanted Christmas

Crawl 5
Valkenburg

Note this town is quite compact and the distances between drinks will be a lot shorter than in previous crawls, so be advised to pace yourself. The good news is that you have more time in each pub! Also as it is really compact, we will fill the gaps with Christmassy bits.

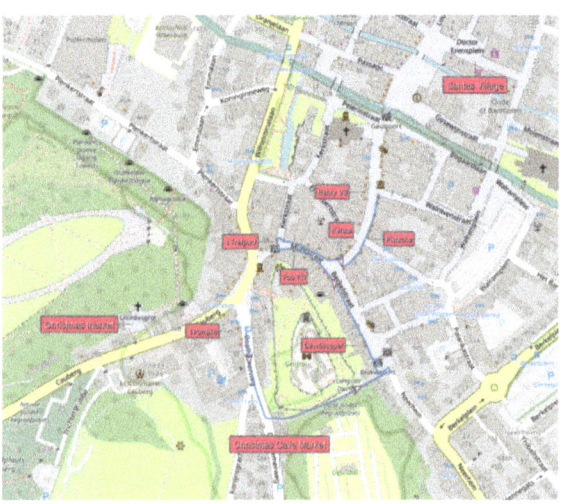

https://www.plotaroute.com/map/1908694

Let's clear up something to start for those who are unsure: When is the Netherlands called Holland?
Here is a potted overview;

- Officially known as the Kingdom of the Netherlands, some people use Holland or Netherlands. They are not the same!
- The Netherlands is actually made up of 12 provinces.
- Holland which is a part of the Netherlands, account for 2 of the 12 provinces, Noord Holland and Zuid Holland.
- The other 10 are Drenthe, Gelderland, Groningen, Flevoland, Friesland, Limburg, Noord-Brabant, Overijssel, and Utrecht, Zeeland.
- Going back in time, between 1588 and 1795, this area was known as the Republic of Seven United Netherlands.
- Their friendly neighbours, known as The French Directory (the governing five member committee in the French First Republic from 1795 to 1799 until it was overthrown by Napoleon Bonaparte), conquered The Netherlands in 1795.
- Napoleon then appointed his brother Louis as king in 1806, thereby turning it into a 'kingdom'.
- As the area of Holland made the biggest contribution to the Kingdom's economy and wealth, it adopted the name for the whole country.
- Even today, the cities of Amsterdam, Rotterdam, and The Hague are in Holland, so people wrongly assume that Holland is the whole country, not just two of 12.

- Limburg, where Valkenburg is situated, is one of the remaining 10 provinces, so it is in The Kingdom of the Netherlands, not Holland.
- Well, that cleared that up.
- On entering Valkenburg, your eyes may glimpse the ruins, hidden behind the houses, high on the hill Kasteelruine Fluweelengrot. (Castle Ruins and Velvet Cave, which we will be touching on later). Underneath the ruins are a maze of passages.

Valkenburg is quite compact when it comes to finding bars and restaurants to include in a bar/pub crawl. However, that said, although you may not cover a great deal of mileage, there is plenty of variety within this crawl, with the added bonus that, depending on the weather, you are not that far away from the next watering hole. What will be different in this crawl is that I will be injecting some titbits, which, although they may be a stone's throw away from the direct route, will provide you with an interesting perspective of Valkenburg and not just the touristy bits.

So let's go.

Christmas Quote:

"For me, the spirit of Christmas means being happy and giving freely. It's a tradition for all the kids in the family to help mom decorate the tree. Christmas is all about family, eating, drinking, and making merry."

– Malaika Arora Khan

Starting point: Gemeentegrot (Municipal Cave Christmas Market)

We will then be 'crawling' to the following pubs/crawls.
https://www.kerstmarktgemeentegrot.nl/en/home

1. Café T'Pumke
   Address: Daalhemerweg 2; 6301 BK Valkenburg Lb
2. The Dewdropper
   Address; Berkelstraat, 6301 CC Valkenburg
3. Pub 13i
   Address: Muntstraat 13, 6301 BW Valkenburg, Netherlands
4. Café T'trefpunt
   Address: Kerkstraat 1; 6301 BX Valkenburg
5. Pub Karma
   Address: Grotestraat Centrum 6, 6301 CX Valkenburg
6. Pistache Cocktail & Streetfood Bar
   Address: Grotestraat Centrum 15, 6301 CV Valkenburg, Netherlands
7. Henry VIII
   Address: a/d Geul, Grotestraat Centrum 24, 6301 CX Valkenburg
   Overall Route is about O.5 miles
   Start Point
   Gemeentegrot (Municipal Cave Christmas Market)

From the middle of November, this 'cave' morphs into Europe's largest underground Christmas market; As Andy Williams once crooned about Christmas 'It's the most wonderful time of the year'! And that is no better expression, to sum up Valkenburg in wintertime. This little-known hidden gem transforms overnight into the Christmas City of the Netherlands. From the moment you arrive, you will experience the spirit of Christmas, which is exemplified at the Municipal caves, where you will be provided with a truly unique Christmas market experience!

Held in the corridors of a real marl cave/quarry (marl is a soft sedimentary rock made up of clay and usually rich in carbonate minerals. When formed as a rock, it is known as marlstone. Normally found in marine or freshwater areas. You may be aware that marl forms the lower parts of the White Cliffs of Dover), which is sited at the foot of Cauberg Hill.

A former chalk quarry, a WWII hiding place and a nuclear fallout shelter.

Once you make your way inside you will find a treasure trove of Christmas stalls, all selling wares to help you prepare and celebrate Christmas. Maybe some libations as well.

NB. The times listed below are a guide as they may change, so you need to check on the website (referenced above) for the most current opening times and dates.

Monday to Friday*	11:00 am to 7:00 pm
Saturday to Sunday*—	10:00 am to 7:00 pm
Exceptions:	
*December 24	10.00 am to 6:00 pm
*December 25	Closed
*December 26	11.00 am to 6:00 pm
*December 30	11.00 am to 6:00 pm (last day of opening

Did You Know?

Every year the town has an artificial Christmas tree, some 14 m tall is certainly a centre point which emphasises the town's tenacity in making sure it lives up to its self-proclaimed 'Christmas Town'.

Route: As you exit the Christmas market onto Cauberg turn left and take the first right into Daalhemerweg and our first watering hole is just down here on your right.

*If you want a quick detour, I have one now for you before we start drinking. Do fancy having a look at a replica of the Roman Catacombs in Rome, right here in Valkenburg, well if you stay left after coming onto Cauberg, and follow around into Plenkertstraat about ¾ of a kilometre down you will come across the Roman Catacombs.

Roman Catacombs

- In this museum, you can descend to view some Roman Graves and some beautiful Burial Chambers.
- They are about 25 meters below the surface.
- Frescos adorn the walls, recalling stories, day-to-day life scenes as well as ancient rituals.

- Found in 1910 by a collaboration of Jan Diepen, and Pierre Cuypers, who was a famous architect of the time.
- They run tours to the catacomb which last about an hour (they have 14 catacombs here) and are quite interesting.

You may not wish to detour so let's not dwell.

> **Christmas Quote:**
>
> "Gifts of time and love are surely the basic ingredients of a truly merry Christmas."
>
> – Peg Bracken

Café T'Pumke

Address: Daalhemerweg 2; 6301 BK Valkenburg Lb
Hours:

Day		Hours
Sunday	—	12:00 pm–2:00 am
Monday	—	Closed
Tuesday	—	Closed
Wednesday	—	6:30 pm–2:00 am
Thursday	—	6:30 pm–2:00 am
Friday	—	4:00 pm–2:00 am
Saturday	—	12:00 pm–2:00 am

It is an excellent place to start our crawl, a lovely little café with a terrace (not a pub in the English sense).

It is a friendly place, and the staff are very welcoming. Not to be missed.

Route: Turn right out of the café and continue until you come to tuning on the left, which is a footpath Van Meijlandstraat. Take this and continue alongside the castle grounds d down the steps onto Neerhem, then turn left under the Castles Arch/Bridge into the pedestrian zone, and the Dewdropper is just down on your left.

Onwards to Velvet Cave Christmas market.

Underneath the ruins of Valkenburg Castle, centuries-old corridors meander through the yellow marlstone. The Christmas market spreads its stalls through the caves each year, making visiting the Velvet Cave (Fluweelengrot) even more enchanting.

There are lots of decorated stalls with unique Christmas ornaments. Check out the charcoal drawings and cave art on the marlstone walls. The aromas of typical Christmas market goodies intensify the inviting, underground catering section.

(Check with the website for exact times etc.).

Castle Ruins & Velvet Cave

- The Netherlands' only hilltop castle towers over Valkenburg; it is now only a ruin but offers fantastic city views. Most of the castle was made of limestone blocks from the local caves. Walking around, you get a real sense of what life was like in 1110.
- Many informative signs are written in English to guide you through the ruins; they describe what was

happening and when it happened, as well as how they built the castle.

- Fantastic artwork hangs on the walls of the caves for you to admire as you explore, and make sure you head to the top and see the fantastic panoramic view of Valkenburg.

Wilhelmina Tower

- Built in 1906, Wilhelmina Tower now allows visitors to take a relaxed journey by chair lift up to the top of the mountain and visit the Tower. The panoramic views are stunning (on a clear day); you may even glimpse the border of Belgium and Germany. Good luck distinguishing one from the other.

Valkenburg grew up as a fortified town at the base of the hill, with the castle on top surrounded by bulwarks, towers and other defensive structures.

Originally the city had seven city gates, but today, only three remain one at Grendelpoort, another at Berkelpoort and the last at Geulport. As you meander, I am sure you will come across them.

Valkenburg Castle

- It appears the first beginning of a fort started here in about 1115 by Lord of Valkenburg, Gosewijn I.

- The fortifications were not substantial, mainly being built of wood, which made them easy for Emperor Hendrik V to destroy in 1122.
- Over the next 200 years, the incumbents made more transformations before being demolished by Jan III (Duke of Brabant) following another siege (is there a pattern here?) in 1329.
- Like the proverbial Pheonix, which rose from the ashes, it was rebuilt again, but by then, the construction methods had improved, meaning that renovation is the castle we still see here today.
- So as with any fortress on a hill, the most efficient way to defeat them is by laying siege to them.
- So over the next 200 years and three or four more sieges, it was finally laid to waste in 1672.
- King-Stadtholder Willem III ordered his troops to blow up the castle to stop it from falling into the hands of the French.
- Here is where the sad tale of the castle slips away into the history books.
- It was never rebuilt and was just allowed to deteriorate into ruins.
- So the next steps in its pathway to where we are today include Napoleon's troops confiscating it in 1795, who then sold it on. A few more ownership changes happened until the Valkenburg Castle Foundation bought it.
- Now jump forward to go back in time. Puzzled? Let me explain.

Just before the start of WWII in 1937, work was being undertaken to renovate parts of the castle, and during that work, a series of not before seen (in recent times) tunnels and passages were discovered. It is believed that these were originally used by Knights and hewn out as escape routes and or for breaking out during the sieges which plagued the history of the castle.

Some of the passages lead to the Velvet Cave and are part of an underground labyrinth, which was used during WWII to hide/shelter people during the fighting.

- At the start of WW II, the Netherlands had intended to remain neutral (as they did in WWI).
- But things changed rapidly as the Nazis invaded and occupied Valkenburg in 1940.
- Not much changed initially, and it was believed by the Nazis that the 'German Dutch' would eventually go along with the new regime.
- However, the resistance movement exploits in Valkenburg have yet to be widely known. After the war, for some reason, a lot of their work should have been spoken about for everyone to know of their exploits.
- I will leave you to delve further if you so wish, but as this is just a brief look, I will whet your whistle with the following information.
- So after four years or so of Nazi rule, in 1944, the allies were now advancing across Europe; the Nazis were doing everything to slow their advance and protect their own retreat.

- So in September, as the allies advanced on Valkenburg, the Germans were entrenched on the other side of the River, separated by the bridge at Wilhelminalaan. This bridge had been kept intact to allow the retreat of their forces.

- On 7 June 1944 (the day after D-day), the first troops of the 19th US Army Corps landed on European Soil, and nearly 13 weeks later, a small section walked into Valkenburg.

- What they found was a ghost town, it was deserted!

- Why had all the people that stayed chosen to hide in the caves?

- As time passed, two men ventured out into the street from the caves.

- These were Pierre Schunck and another guy who hailed from The Hague.

- They provided information to the Yanks about where the Nazis were encamped.

- However, not being very trustworthy, who could blame them? The Yanks were not 100% about the information being legit!

- So, they devised a plan whereby the informer would 'out their life where their mouth was'.

- A line of Jeeps approached the bridge at Wilhelminalann, which crossed the River Geul, with Pierre sitting on the front of the first jeep (with a Yank's gun pointing at the back of his head, just in case it was a set-up)!

- To spoil things, some 'grass' had let the Nazis know an attack was imminent, and the bridge was blown up. They had planted explosives previously for just such an occasion.

- It was not until three days later that the Yanks got the better of them, and the town could emerge from their cave hideout.

Now comes a moment which would never have been foreseen at the time.

Years later, during a wake at the Margraten cemetery, a US soldier started asking around for someone named Paul Simons, but almost everyone had forgotten that name. When he finally found him, it turned out he was the soldier who sat behind him with a rifle pointed at his head. He had had sleepless nights because of this and was happy to find Pierre Schunck in good health. This soldier was Bob Hilleque from Chicago, the only member of the A platoon of the 119th regiment who was still alive at the time. Pierre and Bob subsequently became good friends.

'History of Sorts Forgotten History-Pierre Schunck Resistance Fighter'

Christmas Quote:

"Christmas is a necessity. There has to be at least one day of the year to remind us that we're here for something else besides ourselves."

– Eric Sevareid

The Dewdropper

Address; Berkelstraat, 6301 CC Valkenburg
Hours

Thursday	—	3:30 am–10:00 pm
Friday	—	3:30 pm–12:00 am
Saturday	—	1:30 pm–12:00 am
Sunday	—	1:30 pm–10:00 pm

The opening hours are strange, so you must check if it is open before you go. If not, plenty of other alternatives are nearby while still on your overall route.

Route: As you come out of the Dewdropper, turn left and stay on Berkelstraat for just over 100 m, and the next pub is on your right.

Okay, let's have the trivia stuff about The Netherlands.

In Europe, The Netherlands is the most densely populated. UN Data in 2020 show the population was just over 17 million, and given a land area of over 33000 sq/km, gives a population figure of about 508 people per sq/km.

This means it is Number 1 in Europe and is high in world rankings.

- In Europe, the Netherlands is the lowest-lying.
- Not really a shock as the name the Netherlands actually means 'lower countries', but it may shock you that over a quarter of it lies below sea level.
- The Netherlands ranks sixth in the world among the happiest countries!

In the 2020 World Happiness Report, out of 156 countries (that were in the survey), The Netherlands came sixth. I will leave you to decide whether its neighbours are suitably placed. Germany came 17th, and Belgium came 20th. The happiest places are Finland, Denmark, Switzerland, Iceland, and Norway.

- Regarding Diet, The Dutch are reputed to be the world's healthiest!
- Coupled with being one of the happiest nations, the Dutch also claim to be one of the healthiest. The Oxfam Food Index rates the Netherlands as Number 1 for having the best nutritious and healthy food selection. (and it's very affordable). This makes sense when you realise that the Netherlands is the second largest exporter of veggies.
- The Dutch are also number 1 in Europe for being the most physical.
- If we use the Euro Barometer, about 56% of Dutch people play sports weekly (the Euro average is 40%). The infrastructure and lack of gradient also help to encourage people to do a lot of cycling! This also may explain why there are more bikes in the Netherlands than there are people.
- The Netherlands is proud of its diversity.
- More than three million of the country's residents were born outside of the Netherlands.
- It may be strange, but The Netherlands has only one official language.
- Being so diverse and with lots of people speaking other tongues, English included, Dutch is the only

official language of the Netherlands. However, it should be noted that the Netherlands also has its own Dutch Sign Language (NGT).

- One usually associates windmills with the Netherlands, but why?
  - So the Dutch started their love affair with windmills back about 800 years ago.
  - By the 1800s, over 9,000 windmills were here.
  - Even today, around 1200 still exist, although not all are commercially operational.
  - Some windmills have been used in industrial processes, such as paper, threshing, flour mills etc.
  - However, due to its topography/geography, the Netherlanders built windmills to pump water out of the lowlands and send it back into the rivers beyond the dykes. This then became a form of flood defence.
- Another Dutch iconic item is the Clog.
  - The Dutch have been wearing clogs (or klompen) since medieval times to protect their feet while doing hard labour. Talk about forward thinking. Even back then, they could see the safety issues with some types of work and produced 'personal protective equipment' to reduce accidents/injuries.
  - Why clogs? Well, they are hardwearing, easily cleaned and, best of all, waterproof.

- o   Now mainly confined as a tourist 'take home' item, they still produce in excess of 6 million pairs a year.
- As Max Bygraves crooned 'Tulips from Amsterdam', but are they?
  - o   Believe it or not, the tulip did not originate in The Netherlands.
  - o   In fact, it was brought here in the 1600s from Turkey.
  - o   When they arrived, the Netherlanders developed a passion bordering on obsession with this little bulb.
  - o   So much so that by the 1630s, the prices for these bulbs rocketed, costing almost as much as a house!
  - o   But with all the 'booms', the 'bust' was not far away.
  - o   Due to the profits that could be made, farmers converted their land to the growing of tulips.
  - o   So in 1637, the bubble burst causing financial ruin for many.
  - o   Yet today, they have returned to people's list of favourite plants, and a whole tourist industry has popped up around visiting the tulip fields.
- They like their beer here!
  - o   Famous for many beers, including Heineken and Grolsch, plus others. It stands to reason that the Netherlands is the second biggest exporter of beer, in the world, just behind Mexico.

- Gin, a truly English spirit!
  - In Britain, it is widely recognised that Gin holds a special place in drinking society, but do you know where it was invented?
  - As with a lot of history, there are various claims from people/counties claiming to be the first etc., in many things, and Gin is no different.
  - We do know that the first mention of Jenever (Gin) in writing was in the thirteenth century in Der Naturen Bloeme.
  - The first gin recipe (in print) was from the 1700s and found in Antwerp.
  - But we can go back further to the eleventh century in Salermo, Italy, where a Benedictine Monastery was sited in the middle of an area full of juniper berries.
  - But let's say the Gin we know today started near here but was made popular by William of Orange. (King William III of England etc.) brought it over here, and indeed it became a spirit used to 'gird the loins' of soldiers before going into battle in the 30-year war hence the phrase 'Dutch Courage'.
- As we are talking about William of Orange, let's talk about Orange.
  - You may wonder why, when the flag of the Netherlands has no orange on it, their national colour is Orange.
  - But if you look at their Monarchy, the family is 'The House of Orange'.

- o So basically, the Netherlanders demonstrate their support by wearing the colours of their Monarchy, Orange.
- Is Orange an obsession?
  - o Here is a great piece of trivia, the Dutch turned carrots orange!
  - o For hundreds of years, the carrot had a multitude of colours, white, yellow & purple.
  - o But in the 1600s and as a homage to William, the Prince of Orange, growers developed an 'orange variety' of the carrot. The rest, as they say, is history. It became the go-to colour for carrots.

Christmas Quote:

"Faith is salted and peppered through everything at Christmas. And I love at least one night by the Christmas tree to sing and feel the quiet holiness of that time that's set apart to celebrate love, friendship, and God's gift of the Christ child."

– Amy Grant

Pub 13i

Address: Muntstraat 13, 6301 BW Valkenburg, Netherlands
Hours:

Sunday	—	12:00 pm–2:00 am
Monday	—	12:00 pm–2:00 am
Tuesday	—	12:00 pm–2:00 am

Wednesday	—	12:00 pm–2:00 am
Thursday	—	12:00 pm–2:00 am
Friday	—	12:00 pm–2:00 am
Saturday	—	12:00 pm–2:00 am

This is a modern 'Dutch' pub with a wide range of drinks, food and entertainment.

As with the traditional pubs that most people are aware of, this pub has that but still maintains its Dutch identity. A must-visit!

Route: A shop hop, step and jump, turn left out of the pub and then first right, and there you have it.

Christmassy Bits

- It was the Ancient Egyptians who started decorating pine and fir trees. A tradition carried in households across the world today.
- Santa Claus actually derived his name from the Dutch Sinter Klaas.
- Santa Claus's image is linked to Coca-Cola!
- He used to dress in green (remember the Green Man earlier!), but after Coke decided to dress him up in red and White to match their brand, it stuck!
- Baby Jesus was given Gold, Frankincense, and Myrrh by the Three Kings (known as the Magi), but who were they, and why did they choose these presents? Well, nowadays, one might be forgiven if you did not recognise two of the three gifts, but back in the day,

they had specific uses and or meanings, so; let's take them one at a time.

a. Melchior brought a casket of Gold, now Gold even back then had links to royalty, and all Jesus was the Son of God (King of Heaven).

b. Casper (aka. Gasper) brought Frankincense, which was a perfume/incense and had links to Deities.

c. Finally, Balthasar brought Myrrh; now, this was an anointing oil, normally associated with death.

d. So in their way, they not only celebrated his birth but nigh on foretold his future. Jesus down to Earth from Heaven (as our Saviour [Gold]), Evangelised, gave us the gospels and showed us how to get to his Father's Kingdom, [Frankincense/incense always linked to religious services etc.] and lastly, he died to save us (Myrrh).

• Christmas pudding or Plum pudding! Strangely enough, this Christmas pudding does not contain plums! What you say, no Plums! That's correct, so let me tell you a story.

o The Plum pudding became the go-to Christmas dinner dessert in about 1650.

o However, the do-gooders, back then known as the Puritans, tried to ban it in 1664, as it was their belief that this 'pudding' was overly rich and therefore not suitable for true God-fearing believers.

- But thank God, well, in fact, thank King George I, who in 1714 brought it back into the Christmas meal. This was because he tasted one and was so impressed he sang its praises so much so that by the mid-1700s, it was re-established.

- We all know the song 'The Twelve Days of Christmas', where a lover brings different gifts every day to their sweetheart for 12 days. But why 12?
  - Well, there is a good reason, as we found out about the Three Kings who travelled to see the baby Jesus, but did you know they travelled 12 days to reach him? (This may have been because although Jerusalem is close to Bethlehem. So the Star and the Magi took a longer route to stop soldiers from following them.)
  - The arrival of the Three Kings is also celebrated as the Epiphany; indeed, some countries mark this day with major celebrations.
  - Now on a sidebar, people always discuss when the decorations should come down after Christmas; well, here is my take on that matter.
  - We light the top of our Christmas tree with a Star to symbolise the Star the Three Kings followed, so it stands to reason the Star remains on the tree until they arrive when it's no longer needed to show them on the way,

and that is when the decorations and tree come down; Epiphany Day.

- It was also the Dutch who came up with the tradition of leaving out cookies and milk for Santa.
- Did you know that Christmas presents helped WWII POWs escape?
  - Apparently, in WWII, the Secret Services of the allies sent playing cards to the POWs being held in Nazi camps as Christmas presents. Seeing no harm, the Germans allowed them to be received; however, what they did not know was that when the cards got wet, they split apart, showing maps of escape/safe routes out of Germany! Damned clever, eh!
- Hopping back to England for a bit, did you know that The Holy Days and Fasting Act of 1551 is still on the statutes today, and it states that in Britain, all its citizens have to attend the Christmas Day church services? It also stipulates that no form of transport other than your feet is to be used to get you there and back!
- Another nice little belief is that if you Bake bread on Christmas Eve and it will remain eternally fresh and mould resistant. I guess it would not take too long for believers in this 'miracle' to discover it was not really true; I suggest about 4–6 days would do it.
- Now back to the Yule log. In Scandinavia, the tradition of burning it goes back centuries.

- The Yule log was always seen as a totem of good luck, and thus its remnants (ashes etc.) were kept in an attempt to keep good luck and fortune over the next 12 months.

- But only some people are or were as keen as I in Christmas. In fact, the celebration was deemed against the law in England from 1647–1660. Now whom would you think would have had no qualms in enforcing this law? You guessed it, our good friend Oliver Cromwell. He vehemently believed it was abhorrent and blasphemous to celebrate and have high jinks on one of the holiest days of the year. Therefore it became a criminal offence.

- Now let's talk about gifts again. We can trace back gift giving to Paganism and to Saturnalia, both of which used to show they cared for someone by giving them gifts. However, yet again, do-gooders (the Puritans again!) did not like the idea of Christmas celebrations being tainted with pagan rituals such as gift-giving. But once again, they lost out as those who liked the idea made a comparison to the Gifts of the Magi, and so the tradition stayed.

- Some people cannot wait for the Christmas party at work to grab a kiss under the Mistletoe! But did you know, in some cultures, it was believed that any woman who ventured under the hanging leaves and berries and did not manage to get a kiss would remain single for the next year?

Café T'trefpunt

Address: Kerkstraat 1; 6301 BX Valkenburg
Hours:

Sunday	—	12:00 pm–11:30 pm
Monday	—	Closed
Tuesday	—	Closed
Wednesday	—	12:00 pm–11:30 pm
Thursday	—	11:00 am–11:30 pm
Friday	—	12:00 pm–11:30 pm
Saturday	—	12:00 pm–11:30 pm

Again, and I have to say it's not unusual, but when I was there, you were totally made at ease by the owners who do their best to create a great friendly atmosphere. Lots of people return time and time again to this café. Well worth the stop, even though it's only a short journey from the last.

Route: Another short hop. Go on back down Munstraat keeping to your left, and the next watering hole is on your left after about 100 m.

Okay, a few bits to keep you going before our next pub.

- What comes to mind when I say Christmas and Santa Claus?

  - Santa Claus would be somewhere near the top of the list.
  - He has been associated with Christmas since the Middle Ages, in one shape or another.
  - Although it was not until the 1870s that we first see him portrayed as a jolly fellow dressed in red with a black belt etc.
  - The character of Santa Claus was quickly linked to St Nick (who died way back on 6 December, 343 AD, *NB: that feast day is still held by the Dutch for giving gifts etc.*), a local Bishop who would give gifts to poor children.
  - So how did Santa end up at the North Pole?

    - Well, cartoonist Thomas Nash produced 33 drawings for Harpers Weekly magazine in America, all with Christmas Theme (indeed, these were the first to show Santa in Red etc.).
    - One of these sketches also showed the little town of Santa Claussville, N.P) [N.P. was the abbreviation for the North Pole!]
    - Was this just a 'light bulb' moment? Well, no, not really. To grasp the reasons why you have to think back

to that time. During the mid-1800s, the world was in awe of the Expeditions to the Poles etc., which had not been explored previously.

- Reason 2, remember *The Night Before Christmas* poem, well in that he mentions reindeer as Santa's choice of transport. Can you guess where reindeer live... The North Pole?

- Reason 3. It is covered in snow and ice, and snow is associated with Christmas worldwide.

- Reason 4, Accessibility to the North Pole is not easy, so it would be hard to dispel that Santa lived there.

- According to the Guinness Book (World Records) the biggest snowflake on record fell in January 1887 during a snow storm at Fort Keogh, Montana. The tale of the tape was that it measure about 15" wide & 8" thick!

- Which flower do you associate with Christmas? Is it Poinsettia?

  o Well, back in the 1820s, Mr Joel Poinsett sent samples of the plant back to the United States from Mexico (although he was a United States Minister, he was also a Botanist). They soon became a favourite, and by the 1870s, they became widely associated with Christmas.

224

- Now, what item from Denmark has caused many a parent to jump around the living room? You guessed it, Lego!

  - Buts its popularity is staggering. In 2022, around 220 million Lego sets were sold worldwide; that's about one every seven seconds.

- Now we all know if you write a Christmas song which is successful, you get the royalties year after year, and it can be the making of you.
- But did you know back in 2006, a copy of 'Twas the Night Before Christmas (signed in 1860 by Clement Clarke Moore) was sold at auction for $280,000.
- Talking about Christmas Songs, a little-known fact is that the hit *All I Want for Christmas Is You* was written by Mariah Carey in just 15 minutes.
- How's your cooking? A Food Network survey found that women (nothing about men!) from Britain (on average) do not try a cook a full Christmas dinner with all the trimmings until they are 34!

  - Now is that because they are nervous, or is it because they are partying hard and are recovering? Who knows!

- Now back in the U.K., and probably with a lot of stock from the Netherlands, Brits will drink approximately 57 Olympic-sized swimming pools worth of beer over Christmas; each pool holds approximately 2.5 million litres', which then equates to 142.5 million litres. That's 300,960,000 pints!

- Love or Hate them Brussels sprouts (and the name is Brussels Sprout, not Brussel Sprout!) They are a staple item on the Christmas dinner plate (boiled, broiled, steamed, tossed in butter and bacon; anything to do away with the taste). They're so popular in the U.K. growing them all uses an area of over 3000 football pitches.

- Wayne Sherlock is definitely a lover of the little vegetable. So much so that he attempted to break the world record of eating more than 31 of the little devils in one minute. Whoop, Whoop, he did it, scoffing a sickening 33 in just one minute.

We know from history and Charles Dickens that turkey was the rich person's meat of choice at Christmas, while the less well-off had goose.

If you remember from Dickens' A Christmas Carol, the Cratchit family was settling down to have a roast goose, but that was before Scrooge burst in on them with a massive turkey.

Another strange tradition from Sweden is that they create a giant Yule goat (made of straw) to celebrate Christmas.

Now Scandinavians are not adverse to a goat or two in their myths and legends etc. If you think back to the Old Norse Gods, Thor (and he was my favourite as a kid) used to fly about in his chariot, which was pulled by two goats (similar to Santa and his reindeer). The Yule goat, so the legend goes, drops in on every home, checking everyone is getting prepared for Christmas Day.

Pub Karma

Address: Grotestraat Centrum 6, 6301 CX Valkenburg
Hours:

Sunday	—	12:00 pm–10:00 pm
Monday	—	12:00 pm–10:00 pm
Tuesday	—	12:00 pm–10:00 pm
Wednesday	—	12:00 pm–10:00 pm
Thursday	—	12:00 pm–10:00 pm
Friday	—	12:00 pm–10:00 pm
Saturday	—	12:00 pm–10:00 pm

Another pub restaurant, which has a fantastic selection of beers, and good food. You can even catch a glimpse of the castle from outside.

Route: Turn left out of Karma, and continue down Grotestraat Centrum for about 50 m, and Pistache is on your right.

As we go our way, here are some other important dates in the Netherlands' history.

Before Christ (BC)

- From about 2000 to 800: The Bronze Age folk live in the region.
- 800–58: Well, this was the Iron Age, and during this time, Germanic and Celtic tribes moved in.
- At last, a specific year and wouldn't you know it, it is thanks to the Romans (One could ask, "What have the Romans ever done for us?" (Monty Python joke), but seriously, Julius Caesar invades Southern Netherlands and takes it as part of the Roman Empire!
- Anno Domini (AD)
- 400: Start of the Empire's fall, as the Romans are beaten back by Germanic tribes. This opened to door to other tribes from the Saxons, Franks, Angles, and Jutes to move in and make their home.
- 768: King Charlemagne [aka Charles the Great; born 2 April 747, died 28 January 814. A well-known king throughout history became King of the Franks in 768, then King of the Lombards in 774, and finally was crowned Emperor of Romans in 800] in 768, he decided to expand his Empire to include the Netherlands.
- 800–1000: So, over the next two hundred years, the Vikings decided to come along and invite themselves to some pillaging. Some liked it so much that they stayed.
- 1083: This year, Holland is first used in a legal document. It described an area later recognised as the County of Holland.

- 1384–1482: This is known nowadays as the Burgundian Period. During this period, the Duke of Burgundy united and ruled most of the Netherlands.
- 1482–1567: This time in history saw the end of the Burgundians in the Netherlands and started the rule of the Habsburgs.
- 1568: This year saw William I, Prince of Orange (not to be confused with William III, who, although part of the Orange dynasty and ended up being King of England) lead a revolt against the Habsburgs, and thus started the 80 Years War.
- 1581: A notable date for the Dutch as they declare their independence from Spain. Thus is born the Republic of the Seven United Netherlands.
- 1602: This year, the Dutch East India Company (aka United East India Company) was founded in the present-day Netherlands (but back then, it was the Dutch Republic). It was formed to protect the Dutch's trade in the Indian Ocean and to help (in any way it could) the home nation in the Dutch War of Independence from Spain.
- 1642: This year, Rembrandt painted his famous painting, The Night Watch, which is on display at the Amsterdam Museum.
- But here is a nice bit of trivia for you. Did you know that the title The Night Watch is not actually a night-time painting?
- Believe it or not, it depicts a daytime scene, and the painting was not called The Night Watch by Rembrandt.

- It got the name during the eighteenth century, when it may have been mistakenly thought of as a night-time scene due to the build-up of dirt and varnish.

- It shows a group of civic guardsmen whose role was to serve as defenders of their cities. Tasks included defending the city's gates, keeping law and order in the street as well as extinguishing fires. They also played a significant role in civil parades and VIP visits.

- 1648: This is a memorable date in the history of the Netherlands, as the signing of the Peace of Munster, by the seven United Netherlands and Spain brought an end to the 80 Years War and the recognition of the Netherlands as an independent country.

- 1652: The Dutch have War declared on them by the English following some naval disputes over trade. Both sides had heavy losses, but eventually, the English were the victors (after approximately two years).

- 1795: Another invasion, this time the French, leading to the establishment of the Batavian Republic. This link with France was a milestone in the Netherlands' history as it eventually led to the United Kingdom of the Netherlands.

- 1806: No nepotism here then, as Emperor Napoleon Bonaparte installs his brother, Louis, as King of the Netherlands.

- 1813: Defeat looms for Napoleon and the French. And in this year, the United Kingdom of the Netherlands was declared. At this point, it also incorporates Belgium but has two recognised capital cities: Amsterdam and Brussels.

- 1830: This coalition of countries did not last more than 17 years, and this year, Belgium broke away to form its own country.
- 1853: In this year Vincent van Gogh was born on 30 March, in Zundert, Netherlands, and died aged only 37 on 29 July 1890 in France.
- 1917: A landmark moment in this year, Women of the Netherlands won the right to stand for an election. Actually, gaining full suffrage some two years later, in 1919. They were well ahead of the curve as only Finland and Sweden were ahead of them in Europe by giving their women the right to vote earlier.
- 1939: A sad date in history as World War II breaks on 1 September. The Netherlands tries to remain neutral but to no avail!
- 1940: Some nine months later, on 10 May, the Germans crossed the border and occupied the Netherlands.
- 1941: In the Netherlands, this is the year the Nazis start to round up the Jewish citizens and send them to concentration camps.
- 1944: On 3 September, during the last throes of the occupation, Anne Frank was discovered by the Nazis and sent to Auschwitz on what would be the last transport from that area, having been in hiding for over two years.

- 1945: 2 September World War II came to an end, and The Netherlands joined the United Nations, securing membership on 10 December this year.
- 1948: Following the end of WWII, The International Court of Justice (ICJ) is established in the Peace Palace in The Hague.
- 1949: The Netherlands finally loses its neutrality by becoming a member of NATO this year.

Christmas Quote.

"Christmas is a piece of one's home that one carries in one's heart."

– Freya Stark

Pistache Cocktail and Street Food Bar

Address: Grotestraat Centrum 15, 6301 CV
Hours:

Monday	—	Closed
Tuesday	—	Closed
Wednesday	—	12:00 pm–12:00 am
Thursday	—	12:00 pm–12:00 am
Friday	—	12:00 pm–12:00 am
Saturday	—	12:00 pm–12:00 am
Sunday	—	12:00 pm–12:00 am

This may be getting repetitive, but here is another bar which welcomes you with open arms. (Valkenburg is such a nice place!) Great atmosphere and cocktail are to boot! By its own admission, its burgers are the 'best in town'. As the name suggests, other street food is available and is a good selection of beers.

Route: Turn right out of the pub and continue down Grotestraat Centrum for another short hop and jump (about 50 m), and you can easily see Henry VIII on your right.

Again as it's only a stone's throw away, here are some more Christmas bits you can digest while drinking.

- In Japan, their go-to meal for Christmas Day, believe it or not, is KFC. It is so popular people are told to place their orders at least eight weeks ahead.
- Talk about thinking outside of the box; back in 2010, the authorities in Columbia came up with the idea of covering jungle trees with Christmas lights, so as the

FARC-EP (The Revolutionary Armed Forces of Colombia) guerrillas passed by, the trees automatically lit up and highlighted banners asking them to lay down their arms. That year over 300 terrorists 'saw the light' and came away from the dark side.

- Now we all know that people's tempers can become frayed when alcohol is involved, but if you fancy a dust-up, why not travel to Peru? There is a village there where people settle scores from the year past by putting up their 'dukes'. That way, all grudges are forgotten, and they start the New Year afresh.

- As we are hopping around the globe, a little-known fact is that in Sweden, lots of families settle down on the 24th of December to watch Donald Duck cartoons. This has been happening since 1959. In 1959, at 3 p.m. on Christmas Eve, the 1958 Donald Duck and His Friends Wished You a Merry Christmas was screened on Sweden's main public television channel. This Disney holiday special (hosted by Jiminy Cricket) opens Christmas cards that reveal film clips/cartoons etc. Back in the 1960s and 1970s, it was the only time the Swedish people could watch Disney on TV, and since then, it has been a Christmas staple.

For those that dislike Xmas and think it's somewhat disrespectful; the letter X, as used in Xmas, is the Greek abbreviation for Christ.

In England (and it doesn't happen very often!), for a Christmas to be classified as 'white', a single snowflake needs to be recorded (seen) on the roof of the Met Office HQ in London on Christmas Day. (In the 1900s, there were only seven recorded White Christmases in England.)

Okay, let's look at another countries traditions:

Iceland

- As the days are so short and dark (being one of the lands of the midnight sun), Christmas lights go up quite early.
- They even celebrate Christmas early, as their main day is Christmas Eve.
- Preparing the Christmas meal is a big thing over there, as baking provides an opportunity for families and friends to do something together while the kids dance around the Christmas tree and sing Christmas carols.
- There are 13 Santas, or Yule lads, in Iceland, with each providing children with a small present in their shoe for each of the 13 nights before Christmas. If the kids are naughty, they get an old potato. But for the original story, see below.
- Let me introduce to you Gryla, a giant troll who lives atop a mountain with her 13 boys!
- These Yule Lad's names are thought-provoking on their own.
- Sheep-Cote Clod: He likes to suckle yews down on the farms.
- Gully Gawk: He is a bit fussy; his thing is sipping the foam from freshly drawn cow's milk.

- Stubby: As you might guess, he is a tad small and 'nicks' food from frying pans.
- Spoon Licker: Can you guess what he does?
- Pot Licker (aka Pot Scraper): This lad likes to help out by stealing unwashed pots/pans and licking them clean.
- Bowl Licker: Unlike his brother, rather than licking, this lad steals bowls of food.
- Door Slammer: He is a torment & keeps people awake by stomping about slamming doors.
- Skyr Gobbler: This lad has a penchant for the Icelandic yoghurt (Skyr).
- Sausage Swiper: As the name suggests, he loves to acquire sausages.
- Window Peeper: Not a pleasant chap creeping around and peeping through windows to see if anything is worth stealing.
- Door Sniffer: Easily distinguished by his very large hooter, and has a huge appetite for food, especially baked goods.
- Meat Hook: Beware this fellow; he will snap up any meat which is left out and is very partial to smoked lamb.
- Candle Beggar: In the past, when candles were a very important part of Icelandic life, this lad liked to nick any candles lying about.
- Gryla's on hubby No.3, having popped off the first two for being boring.

- She was renowned for capturing naughty children, with the aid of her 'lads' and boiling them up alive to make her favourite stew.
- They have a giant black cat, known as Jolakotturinn, the Yule Cat (strangely, Jola is the name of the Yule Cat in the film *Christmas Chronicles 2*), as their pet. Now Jola (for short) does not distinguish over taste. He will eat children and adults alike and doesn't care if you've been naughty or nice.
- But in a twisted way, like Father Christmas (who comes once a year), Jola can only eat once a year at Christmas time. Maybe that explains why he does not care who or why he eats people.
- But he has his own version of Kryptonite; to prevent being an entrée for Jola, all you need to do is receive a new piece of clothing.
- So the only way to avoid being a Yule Cat meal is to get a new piece of clothing at Christmas.
- This story was so terrifying to Icelandic children that the government, in 1746, had to ban using the story of Gryla and the Yule Lads being used as an intimidation tactic to make children behave.

Christmas Quote:

"Christmas, my child is love in action. Every time we love, every time we give, it's Christmas."

– Dale Evans Rogers

Henry VIII

Address: a/d Geul, Grotestraat Centrum 24, 6301 CX Valkenburg

Hours:

Monday	—	11:00 am–1:00 am
Tuesday	—	4:00 pm–1:00 am
Wednesday	—	4:00 pm–1:00 am
Thursday	—	4:00 pm–1:00 am
Friday	—	11:00 am–2:00 am
Saturday	—	11:00 am–2:00 am
Sunday	—	11:00 am–1:00 am

Had to end with an English Connection, and this is good. The inside is like walking into a typical English pub, but the atmosphere which has been present throughout this crawl has not abandoned this pub and is definitely a lot more hospitable than Henry (ask his wives!)

Strange to find a pub with Henry's name in Valkenburg?

Not really, as Henry had a significant relationship with the Netherlands, Antwerp etc.

Henry VIII, at the time, did not need to chase other trade deals as he was more than happy with his dealings with Antwerp and content to follow the arrangements his father had made.

So there is a link between Henry VIII and the Netherlands.

So in the absence of some other history, let's look at some interesting facts about Henry VIII.

- Was he fat?

- In fact, with the average height of a man around that time being 5'6" to 5'8", Henry was actually 6'2" and quite athletic when he was younger, however at the end and with a waist of about 54 inches and tipping the scales at about 320 pounds he was pretty rotund! In fact, he needed mechanical help getting on and off a horse as well as requiring some assistance in the bedroom!
- He was obsessive about his ailments and was practically a hypochondriac.
- Believing he knew best, he self-diagnosed such an array of ailments it is nearly impossible to keep track of all of them. He even thought of himself to be a self-made doctor, prescribing his own medication. Indeed he developed his own prescription book, which can be seen in the British Museum.
- Some believe his last words could have been "Monks! Monks! Monks!" which is quite ironic if true.
- On his deathbed, he suffered from hallucinations, and while the murmurings during his final moments may have been monks, monks, monks, and when you think it was Henry VIII who oversaw the 'Dissolution of the Monasteries' in 1536, maybe he was pleading for forgiveness, or perhaps he was cursing them, we will never know!
- Henry loved to hoard.

His tenure as monarch led England into debt due to his spendthrift attitude and his belief it was his right to spend what he liked. It is alleged that at his demise, he had a personal inventory of 50 Palaces, over 6000 handguns, 70 ships, plus lots and lots of musical instruments, and lots, lots more.

As we leave our last pub, you may want to cross the River to pop into Santa's Village.

Santa's Village Valkenburg

Santa's Village Valkenburg is another Christmas market. Christmas chalets full of excellent local stuff and gifts, along with some warm waffles, crêpes and churros. For a bigger appetite, you can go to Santa's Grill for an original German bratwurst straight from the grill... hmmm, delicious. Then visit our Santa's Bar, where you can relax while enjoying a hot chocolate or original Glühwein.

This is a lovely spot to end the crawl and maybe start your Christmas Shopping.

Christmas Quote:

"You can give without loving, but you cannot love without giving."

– Victor Hugo
Les Misérables

Crawl 6
Ljubljana (Slovenia)

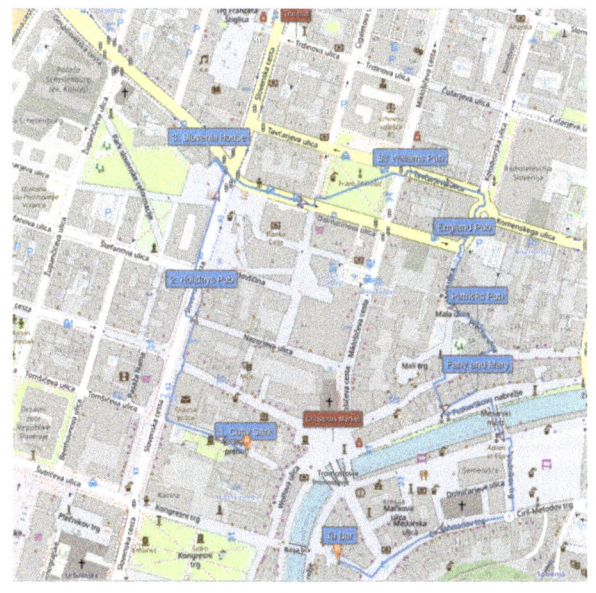

https://www.plotaroute.com/routeplanner

Pubs in this crawl: 1.15 miles

1. Cutty Sark
2. Holidays Pub

3. Slovenia House (something to eat!)
4. Sir Williams Pub
5. England Pub
6. Patricks Irish Pub
7. Fany and Mary
8. Ta Bar

Christmas Quote:

"You know you're getting old when Santa starts looking younger."

– Robert Paul

With a population of approximately 60% Christian, it's hardly surprising that different parts of the city celebrate the festive period differently. Some choose to have their main celebrations on New Year's Eve, followed by a family meal on 2 January.

But as this is Christmas focused, let's see what their traditions are; across other parts of the world, the Christmas dinner is an integral part of the celebration. Now we may be used to turkey and the trimmings, but here the main meal is rolled roast pork, potatoes and sauerkraut; alternatively, you could opt for roast goose and red cabbage with baked potatoes.

But what about presents?

Well, the kids here can start their haul of presents on St Nicolas Day (6 December) from St Nick himself, but it doesn't stop there; some get presents from Baby Jesus, Grandfather Frost and even Santa.

So

6 December: St Nick

24/25 December: Santa Claus or the Baby Jesus and

31 December/1 January: Grandfather Frost.

Ljubljana is such a great city for getting around on foot; the bridges are something else, from the Dragon Bridge to the Cobblers Bridge to the Triple Bridge, which was designed and constructed by Joze Plecnik. (Famous architect who has worked across Prague, Vienna and Ljubljana, you may also read of him elsewhere in the book).

Pinpointing where you are:

Standing on Triple Bridge, you are approximately 200 miles south of Munich (Germany), nearly 300 miles east of Zurich, 220 miles southwest of Vienna, and another 30 miles on top of Budapest. However, you are also 970 miles from London, 4,219 miles from New York and nearly 10,000 to Canberra.

Christmas Market

Held in the Preseren Square by the Triple Bridge just outside the pink salmon Franciscan Church of the Annunciation. Built circa 1646, and red in colour in line with the symbolism of the Franciscan monasteries.

Ljubljana and Slovenia are very proud of being 'green' and promoting sustainability.

Apart from the market stalls and celebrations, other features have been included, such as;

- Magical Forest: this includes interactive workshops which show you how to create your very own decorations which will not damage the Earth.
- The Land of Ice: As the name suggests, artists, both local and worldwide, will dazzle you with sculptures made from snow and ice.
- Organ Grinders: The atmosphere is further enhanced, with Organ Grinders providing a magical auditory background to the wonder of Christmas.
- Music: During the weeks building up to Christmas, you do not need to walk far, for you to stumble upon free street concerts (you could also check local websites too, but where is the fun in that).

*A word to the wise, be careful not to be duped into thinking the classic local aperitif 'Honey Schnapps' is lightweight, it may be smooth, moreish and draws you into its comforting web, but it is very strong!

So as we move across the square to our first pub, remember this is a marathon and not a sprint, don't get too merry too quickly.

Christmas Quote:

"The spirit of the season never has to end as long as you keep your loved ones near, and the holiday wish in your heart."

– Jiminy Cricket
Holiday Wishes

As we leave the market for the Cutty Sark, pause for a moment.

Ljubljana is a lovely city, and it is straightforward getting around on foot. All of its bridges tell a story, from the Triple Bridge in front of you to the Butchers Bridge, the Cobblers Bridge and the Dragon Bridge.

As we face the Triple Bridge, this was dreamt up and built by Joze Plecnik.

There is a reference to this bridge being here since the thirteenth century and was known by various names. The bridges we see here today range from the mid-1800s, and the two which accompany it (this was done to ensure traffic would not cause a jam), were erected in the 1920s. Now all three are from part of the pedestrian zone in the city.

In fact, the bridges form a link from the medieval town to the newer parts of the city.

NB: This crawl will be slightly more Christmas than History, as the city itself is a very 'walkable' city, and therefore the opportunity to highlight things on the route is reduced but don't fear, Christmas will come to your aid!

Cutty Sark

Address: Knafljev prehod 1, Ljubljana

Phone: +386(0)51 686 209, 08 385 91 44

Hours:

Monday and Tuesday	8:00 am–1:00 pm
Wednesday–Saturday	8:00 am–3:00 pm
Sundays and Holidays	9:00 am–1:00 pm

This is a no-brainer, and those of you who have read my previous books, know I have an affinity with the ship 'The Cutty Sark', so seeing this here meant it had to be the first one on my list, and I was not disappointed. Inside and outside areas (undercover) contribute to a fair-sized pub. The décor (like my own bar) is an eclectic mix of ephemera which suits the bar completely.

Route: Turn left when you come out of the pub and follow the path until you reach the junction with Slovenska Cesta. Here you turn right and stay on this road until you reach Holidays Pub on your right-hand side.

As we make our way to the next venue when walking down Slovenska Cesta, on your left three blocks down is St Cyril's Church and Tivoli Park, so let's talk about them here.

Christmas Quote:

"Some Christmas tree ornaments do more than glitter and glow, they represent a gift of love given a long time ago."

– Tom Bake

St Cyril's and St Methodius Orthodox Church

A nice church which is worth a visit, especially if you are diverting to see Tivoli Park.

Why have I mentioned it here, well apart from being my namesake, the church drew my thoughts back again to Prague.

Spookily, there is another church with the same name in Prague (Cyril is obviously a really popular name), but the church in Prague has a bloody history. At 10:30 am on 27 May 1942, an attack on Reinhard Heydrich (Butcher of Prague), from which he later died, took place. The assassins escaped and hid in St Cyrils but were eventually found and killed by the German army.

The story is fully retold in St Cyril's in Prague and was eventually made into a film called *Operation Anthropoid*.

As a side note:

Cyril and Methodius were two brothers who were born in Greece (Thessalonica) in the ninth century. As they were Christian missionaries, their work was based on converting the locals to the ways of God, and they became known as the 'Apostles to the Slavs'.

Christmas Carols

As we have now seen two churches, and it is Christmas, here are some facts about Christmas carols and carolling.

Far from having their origins in Christianity, Carolling was born out of songs which were used to herald the Winter Solstice as part of pagan rituals of the time.

So as the Winter Solstice (the Shortest Day) occurs on or around the 21st/22nd of December (in the Northern Hemisphere).

So as Christianity took hold around the world, the church in Rome declared that Jesus was born on 25 December. Indeed it first appears on a Roman calendar in AD 336.

It is thought this date was purposefully chosen to coincide with the pagan festival Saturnalia as it would be easier to convert the Roman pagans to Christianity and accept it as the official religion of the Empire.

Slowly, slowly catchy, Monkey! This is exactly what the church did, and bit by bit, they substituted/integrated their beliefs etc., into the subtext; consequently, after a while, it became known as Christmas around the Western World, replacing existing festivals.

So as Christmas usurped the pagan winter celebrations, the church provided 'carols' (although this term would not be used for a few more centuries) to replace the pagan songs.

Although probably not recognisable as the carols we know today (they were, after all, performed in Latin), they soon were used throughout the Western World. One of the earliest known 'carols' was Gloria in Excelsis (aka; Hymn of the Angels).

Now jump forward a few hundred years, and things started to heat up!

Early in the thirteenth century, Francis of Assisi started staging little plays about the Birth of Christ (the birth of the Nativity play!)

At the time, it was usual for the story to be told through song (which was predominately Latin) but translated into the tongue of where the play was being staged.

These new 'carols' spread very quickly.

But it was in the fifteenth century that England finally got English-written Carols.

The change was afoot, but as with all our history, things are always challenging. Indeed there was a period of about 200 years when Christmas and carols were allowed to develop organically that was until the early seventeenth century.

During this time in England, the Puritan Movement was gaining momentum, and they thought Christmas had lost its meaning and strived for a more solemn period of reflection rather than the festival, which gave rise to carnal and sensual delights! (I wish I was about before the Puritan came to power, as it sounds like a wonderful time, especially around Christmas!)

But then came along Oliver Cromwell.

Now he may have done a good job of destroying the Government of the Day and killing Charles I (that's another story in another book), but he will also be remembered for 'killing Christmas'.

As Lord Protector, he passed harsher laws which made The Lord's Day a day of sombre reflection, shops were made to open, and the army in London was used to police the area making sure no one was preparing victuals etc., for celebrations.

However, not to be beaten, an underground movement evolved, where masses and Carolling continued in secret. They may have even used the *Priest Holes, which were built in Catholic homes during Elisabeth's I reign, to prevent discovery.

*Priest Hole: A hiding place in plain sight, but disguised to look like part of the wall, chimney breast etc., which could be used by priests at short notice should their services be interrupted etc.

But as soon as the shackles from the Puritanical reign fell away in 1660, Christmas came back with a bang.

Carolling in public really took hold again in the nineteenth century.

Remember here 'The Man Who Invented Christmas' Charles Dickens published A Christmas Carol in 1843.

And Queen Victoria also enjoyed Christmas until the death of her husband, Prince Albert; in fact, they are credited for bringing over the Christmas tree to England.

So What Is a Christmas Carol?

A Christmas carol is defined as a song or hymn which has at its core the theme of Christmas and is played in and around the festive period.

So in effect, any song which has Christmas as its theme could be described as a 'carol', which opens up the whole subject. I suppose you could say if you are trying to compartmentalise it as a genre, Carols would be a subset of Christmas music, but then again, all Christmas music can be classed as carols... that's clear then! Whatever the semantics, here are some facts about Christmas Carols and Music, but in this section, let's stick to the religious Carols.

- 'O Come, O Come Emmanuel', which is still prevalent today, is one of the oldest Christmas Carols, and although written in the twelfth century in Latin, it was transposed into English in the mid-nineteenth century by J.M. Neale.
- Did you know the second song ever played on the radio in 1906 was a violin solo of 'O Holy Night'?
- Silent Night, according to Country Radio Station WYRK, is the most-recorded Christmas song of all time, and in 2017, it was recorded 137,315 times.
- They go back a long way... way before the music charts; in fact, two of the Carols we still sing today were originally penned in English in the eighteenth

century; they are 'Joy to the World' and 'Hark! The Herald Angels Sing'.

- 'Silent Night' was the catalyst for an impromptu truce During World War I.

It was 24 December 1914 when the British soldiers saw Christmas trees lined up in the German trenches. (This scene is brilliantly portrayed in a film called *Joyeux Noel*, which was made in 2005). Then through the still cold night air, they could hear the German soldiers singing 'Stille Nacht'.

Tentatively they joined in using their native tongue. Slowly and nervously, they climbed out of the trenches. Momentum took over, as I suppose the guys were weary, and the thought of even a brief respite was something to be grasped at. They met in No Man's Land, even exchanging gifts and life stories as well as playing a football match (I am not sure if this is the reason why, but there is a long-standing recognition that when we play the Germans at football, it is always a bit special!). Sadly as we know, this ceasefire was short-lived, and in fact, hostilities resumed on Boxing Day. L There was another in 1915, apparently, but it was much smaller and only came to light when a soldier's diary was recovered.

Tivoli Park

This is the biggest park in Ljubljana (it even has a zoo!) and has been a favourite spot since 1813.

Designed by Jean Blanchard and was created by merging the existing parks which encircled the Cekin and Podturn estates.

Covering nearly 5 sq. km. it is a wonderful place to stroll. Lots of tree-lined avenues, fountains and even a place to have a drink and eat while overlooking a lily pond.

Okay, back to the pubs.

Holidays Pub

Address: Slovenska cesta 36, 1000 Ljubljana, Slovenia
Phone: +386 51 290 999
Hours:

Monday	—	06:00 am–03:00 pm
Tuesday	—	06:00 am–03:00 pm
Wednesday	—	06:00 am–03:00 pm
Thursday	—	06:00 am–03:00 pm
Friday	—	06:00 am–04:00 pm
Saturday	—	08:00 am–04:00 pm
Sunday	—	10:00 am–03:00 pm

This has been around since 1992, making this the oldest pub in the city. They also have a place in the city's history. During the War of Independence, the BBC broadcast from the basement, reporting on the issues they faced. But they had to wait till 2015 before they moved outside and provided outdoor seating and tables for our clientele. They now stock a wide range of beers of whiskies, as well as serving food.

Route: Come out of the pub and, turn right, continue along Slovenska Cesta. Our next stop will be a little way down on your left side (NB. It's on the corner).

As there is only a short hop to the next pub and there is nothing much of note along this route, let me tell you about Lake Bled. This is a must-visit if you have the time. It is truly picturesque and once seen, it is easy to understand why people from near and far see it.

Its unique shape came about when the Bohinj glacier created a tectonic hole over the years. The glacier's movement was this humungous rock, which eventually became Bled Island as the glacier's ice melted to make the lake.

The church on the island is called St Mary's Church and has a huge 99-step stone staircase which takes you to the top where you can, if you wish, ring the Bell.

Now I don't know what it is about churches called St Marys and Steps. The St Mary's Church in Whitby, England (of Dracula fame) also has lots of steps, in fact, 199! Why end at 99 or 199?

Bled: The Legend of the Sunken Bell of Lake Bled Aka, The Wishing Bell

There are different versions of the story with slight variances, but whether it was Lady Poliksena, or another Lady living in Bled Castle, their husband was robbed and thrown into the lake, where he drowned. After this, the Lady gathered all precious metals in her home and commissioned a bell to be placed in the chapel on Bled Island.

However, another twist occurred when a storm hit the sea where the boat was travelling with the Bell. All were lost, never to be seen again. Distraught and fed up, the Lady sold all her belongings and moved to Rome, where she entered a monastery and stayed there until her death.

The Pope, moved by her devotion, had a new bell struck, and sent to Bled Island, where it can still be tolled today.

If fact, if you manage to get to the top of the stairs (and there is a lot! In fact, there is a defibrillator at the top, which tells you something about the climb) and strike it three times, you may make a wish.

There is also a spooky side to this in that the sunken Bell can be heard chiming in lost lament on clear nights when the lake is deadly calm.

Christmas Quote:

"Christmas is a time when you get homesick—even when you're home."

– Carol Nelson

So let's go get something to eat and drink!

Slovenia House

Address: Cankarjevo nabrežje 13 1000 Ljubljana
Phone: +386 (0)8 389 98 11
Hours:
Monday–Sunday and public holidays   —
8:00 am–1:00 pm

Slovenian House is a lovely restaurant selling authentic Slovenian Food and drink. A great place to visit, 'cheap as chips' as well, portions are amazing, the staff were very friendly and made out visit an experience. I would have to say of all the places I visited, for food, this was the best!

Route: Come out of the pub/restaurant and cross over Sloevnska Cesta, then turn down (right) into Dalmatinova Ulica. You can continue down here until you reach Miklosiceva Cesta, where you turn right, or you can cut through the park, where you end up (both routes) at the corner of Tavcarjeva Ulica and Miklosiceva Cesta and Sir Williams is in front of you.

Christmas Quote:

"Have yourself a merry little Christmas. Let your heart be light. From now on your troubles will be out of sight."

– Judy Garland

As we leave the restaurant, if you fancy another quick diversion to see a bull, carry on down Sloevnska Cesta, and at the second junction, it will be right in front of you. It's shown on the map above.

Now Ljubljana and the Bull!

If you go to Google and search for 'Ljubljana bull satellite image', you will see an image that overlays a bull's body and a picture of a lily. Why is that important? Well, there are some statues/fountains of bulls in Ljubljana but none more impressive than the one found on the corner of Sloevnska Cesta and Trdinova Ulica.

One wonders if Jakov Brdar, the sculptor had that in his mind when he came up with the idea of a sculpture, I bet he did.

Also, if you look at the lily, it is said to represent the redbud lily flower, which grows on Castle Hill and is a protected species.

Again not a lot along this route, so how about Christmas facts from around the world?

The Pooping Log, Spain

Now when reading this, you may reflect that it sounds like something else you are aware of, which I will reveal at the end of this bit on Spain.

Catalonia, a region in Spain, has some quirky traditions; one such is that of the Tió de Nadal (the Christmas Log). A very easily made decoration which is basically a log with a face and a hat!

Every year the Christmas logs leave the forests and turn up at the children's homes (erm! I wonder how they got there!) on the 8th of December.

Children are then instructed to look after their logs and keep him/her warm and fed (usually orange peel or bread).

Finally, on Christmas Day, the kids gather around with their logs, sing them songs, and hit them with sticks to make them poop out sweets!

Did you think this sounds a lot like a Piñata?

Well, let's look at what a piñata is and its history.

- There is some debate about where it all started; some people swear piñatas go back to Marco Polo's time (circa thirteen century), while others believe it can be traced back even further.

- The Marco Polo camp believes that piñatas were brought back from the East following one of his trips to China and the Far East.

- It was here he allegedly saw little clay-formed animals that were stuffed full of seeds. Like modern piñatas, these were then hit and smashed as part of the agricultural festivities, almost like a sacrifice, hoping that doing this would bring a good crop yield.

- So now let's leap ahead in history to February 1517; it was then that Francisco Hernández de Córdoba, who was the first European explorer to set foot in Mexico. Here, he discovered the Aztecs used clay pots decorated with feathers and filled with small trinkets. Again these were used to celebrate the birth of their god of Sun and War, Huitzilopochtli. As with the earlier story, they were bashed and broken to expose the gifts inside.

- Jump to the present, and this ancient tradition has morphed into donkeys filled with sweets and used at Birthday parties or at Christmas, during Las Posadas (translated as The Inns).

- This religious celebration commemorates the trek Mary and Joseph made to Bethlehem on a donkey. Since that time, the donkey has held a special place in people's hearts.

- The trek from Nazareth to Bethlehem was about 90 miles, so if they travelled eight hours a day travelling at 2½ mph, it would take about four or so days, so you can see why the donkey is a hero.

- It's not much of a stretch to see how the breaking of the piñatas is linked, as the donkey must have really been knackered after that ride.

Scandinavia

A dish of porridge for the Nisse

- A Nisse is a small being of Nordic legend and was associated with Christmas and the Winter Solstice.

- Links can be made to Nisse being like Santa in appearance, if not in size.

- Perceived as being short and stout, sporting a white beard and wearing a red and white funnel-shaped hat.

- So where we generally leave out cookies/biscuits, carrots, milk etc., in colder climates like Norway, they leave out porridge.

- In other Scandinavian areas, there is a belief that every house has its own 'house elf' (where have we heard that before… Dobby!) hiding from sight but is there protecting the family and home. Known as Nisse, it's considered as a 'thank you' to leave them out some porridge on Christmas Eve for all their hard work and protection they have given over the year.

- Another not so well know the tradition of 'Hiding the Broomstick'.

- Not very well known, but it does appear in some texts and stems from the pagan belief witches beset on causing mayhem to come out on Christmas Eve, looking for their preferred mode of transport, so people thought that hiding their broomsticks would limit the damage etc.

- **** Squeamish Alert**** (sorry, but that's just personal to me!)

- Another tradition in Norway is that of Smalahov, a traditional dish made of a sheep's head and served around Christmas Time.

  ○ Each person is provided with half of a head. Normally there is a standard way of eating this delicacy, first of all, the ear and eye are taken first, as being the fattiest bits they are recommended to be eaten while warm.

  ○ They then work from the front of the head round to the rear, digging in between the bones. I cannot find any reference as to whether or not sucking on the bones is considered to be part of it.

- A more well-known tradition (well similar) is the one carried out in Bergen, Norway. That is the Building of gingerbread houses. But Bergen takes it to a whole new level.
- Pepperkakebyen started in back 1991 when it was a school Christmas project to try and make gingerbread models which resembled landmarks, local houses etc.
- It is now reputedly known as the greatest Gingerbread city in the world.
- Any profits from it are donated to local Children's Charities.

Christmas Quote:

"There's a certain magic that comes with the very first snow. For when the first snow is also a Christmas snow, well, something wonderful is bound to happen."

– Frosty the Snowman

Sir Williams Pub

Address: Tavčarjeva ulica 8a, 1101 Ljubljana, Slovenia
Phone: +386 599 44825
Hours:

Tuesday	—	8:00 am–1:00 am
Wednesday	—	8:00 am–1:00 am
Thursday	—	8:00 am–1:00 am
Friday	—	8:00 am–1:00 am
Saturday	—	5:00 pm–1:00 am
Sunday	—	5:00 pm–1:00 am
Monday	—	8:00 am–1:00 am

One of the best places I visited in Ljubljana. The owner was amiable and proud of his establishment, which showed. He had an excellent knowledge of beers and the city. They hold a wide selection of craft beers and other adult tipples, as would be expected, but you should try the Slovenian Gin

(Brin). It has a lovely atmosphere and is just around the corner from the Main Square and River.

Route: Turn right out of the pub and continue down Tavčarjeva Ulica. When you reach the junction of Kolodvorska Ulica, you turn right and continue on this road until you see the England Pub on your right side.

Christmas Quote:

"The rooms were very still while the pages were softly turned and the winter sunshine crept in to touch the bright heads and serious faces with a Christmas greeting."

– Louisa May Alcott

Some more traditions from around the world.

Czech Republic and Slovakia – The Golden Pig

Like other countries, the Czech Republic and Slovakia celebrate Christmas by giving presents on Christmas Eve night.

They hold a family gathering, and after filling themselves with dinner, they share and open presents.

As a 'carrot' to encourage children to finish their Christmas dinner, families tend to fast from meals all day until the feast, (except for goodies). The deal is sweetened for children by the chance of seeing the Golden Pig if they stick to their fast (the sight of the Golden Pig was meant to bring good luck and happiness, and maybe a gift).

As you can imagine, local entrepreneurs have caught on. Now you can buy a range of Golden Pigs, from ornaments to chocolate.

United States of America

Now it's not a massive shock that a country such as America, which has built itself up and developed over the last few hundred years, encompassing a wide range of different peoples and cultures (which in some places have morphed into new cultures with new traditions) that there may be one or two unusual traditions around Christmas.

Okay, here is one which blew me away, and I never saw any link, much like the Spanish 'crapper'. The Gherkin! Or, to put it another way, as the Americans call it, the 'pickle'.

Of course, everyone knows that pickles and Christmas are synonymous with one another, as much as the birth of Jesus, so long as you are American.

Have you ever seen a pickle-shaped ornament on a Christmas tree? Lucky you!

So how did the lowly pickle get elevated to the high position of being A Christmas Ornament?

Let's have a look.

Tradition states that a pickle is hidden among the fronds on the Christmas tree. Whichever child finds it first gets the right to open their present first, or they receive a 'special' gift from the pickle.

Sidebar: Have you ever read Charles M Shultz? Snoopy and Charlie Brown, in his books, has a vegetable giving out presents (supposedly) on All Souls Eve (Halloween)! His one is the Great Pumpkin!

This 'green' tradition (Weihnachtsgurke, which translates as Christmas Pickle) would suggest the roots (pardon the pun) lie in Germany. However, finding any reference to this in Germany or Europe is complicated.

The Christmas Pickle is, however, more prevalent in the states which surround Iowa (mid-west).

Not being presumptuous or pretentious, a small town in Michigan called Berrien Springs has bestowed on itself the title of the Christmas Pickle Capital of the World.

It all starts at the beginning of December, with a parade through the town. The front and centre of the parade is the Grand Dill-Meister (it can be a man or a woman).

But how this yuletide superstition came about is clouded in controversy, as you may have guessed.

There are three popular versions, which could be one of three (or all).

Back in the time of St Nicholas, who lived in Myra, which is in modern-day Turkiye, he came across two children being held against their will by a nasty innkeeper who had locked them in a pickle barrel. As you would expect, St Nick stood up to the plate and released them—that's Version One.

Version Two: During the American Civil War (1861–1865), a soldier of German descent was being held prisoner in Georgia, starving he begged for a 'pickle' (as you would!); his wardens succumbed and gave him his wish, which went on to sustain him until his release.

Version Three: In the mid-nineteenth century, Germany started producing glass baubles shaped like fruit. Woolworths saw this, started buying them, and created a 'market' for them in the USA (clever marketing?)

Russia: Christmas Fortune Tellers

In Russia, things are slightly different amongst the Orthodox community, as they still use the Julian calendar. Their dates differ from the Western world, which moved to the Gregorian calendar.

So 6 January, our Epiphany, is their Christmas Eve, Christmas Day is then the 7th, and their Epiphany then is the 19th of January, which coincides with their belief that this was the date Christ was baptised by John the Baptist.

So with all this in mind, they believe this period to be the best time of year for predicting the future.

So it's no real surprise, then, that 'Fortune Telling' comes to the fore during this time; this can be done in a DIY manner.

This then appeals to unattached ladies, who are especially keen on discovering if the New Year has any good news for them, especially around love and marriage.

But how do they predict the future, not tea leaves of crystal balls, but via mirrors and shadows and other such mediums.

Just a quickie, how do you think they also celebrate their Epiphany... ice cold baths as a homage to Christ's baptism.

Portugal

Bolo-Rei (King's Cakes)

The festive cake (not unlike our Christmas puddings) was baked with a couple of surprises inside, and the outside was colourfully decorated, using fruit etc.

Apparently, this type of cake was very popular in The Sun Kings Courts (Louis XIV 1638–1715) but didn't really embed itself in Portuguese culture until the 1800s.

- During the Sun King's reign, they ate cake at Christmas time, which was known as Cake of the Kings (Gateaux des Rois—a precursor to the Bolo-Rei).
- The French did not wholly invent this tradition; it was an adaptation from a pagan idea, which had a 'party cake' containing a broad bean. Whoever discovered the bean was crowned King of the Party.
- The French decided to go one step further, adding a slight variation to the prize. The person who found the bean would be crowned first King of the Wise Men and had the privilege of laying the first present at baby Jesus's feet on Epiphany Day.
- Not to be outdone, the Portuguese added their own twist. Whoever managed to find the bean would not only be King or Queen for the day but also have the dubious pleasure of paying for next year's Bolo-Rei.

England Pub

Address: Mala ulica 8, 1000 Ljubljana, Slovenia
Phone: +92 305 6739143
Hours:

Monday	—	7:00 am–12:00 am
Tuesday	—	7:00 am–12:00 am
Wednesday	—	7:00 am–12:00 am
Thursday	—	7:00 am–12:00 am
Friday	—	7:00 am–12:00 am
Saturday	—	7:00 am–12:00 am
Sunday	—	9:00 am–11:00 pm

A great little find, if you want a piece of England. A nice addition to the crawl, providing a different atmosphere, and while not a 100% old-school English pub, it's still well worth a visit.

Route: Turn left out of the pub and travel down Mala Ulica. Keep left into Precna Ulica, and Patrick's Irish Pub is just on your left.

As it is only a short hop to the next pub, here are a few facts about the city you may or may not know.

Now known as a Green City, you may be surprised to hear it was not always so.

- The city wasn't always so green, but you have to go back a bit.
  - Long time after the Romans, the first public baths to open in Ljubljana was in 1260.
  - Memorable dates include; 1625; this year, the city rulers decreed that all refuse had to be disposed of 'outside of the city' twice a week.
  - Even later, presumably after it became a problem, a decree was made to ensure all pigs remained within the confines of a sty.
- Ljubljana has been known to have earthquakes, but you could look at this two ways! The last significant one was on 14 April 1895, so it was too long ago to panic… or, It could be a due another one!
- The Philharmonic Hall holds onto a little-known secret. Within its archives is a copy of one of Beethoven's symphonies, penned in 1808! Can you guess which symphony? It was his sixth, aka The Pastoral Symphony.
- Embassies! This here city is the only Capital city where the Russian and American embassies are opposite one another, mind you I would not be taking any snapshots in the area, as you don't know who you will piss off .

- Why an elephant is remembered!
- In Barcelona, there is the Cat; in London, there are the Lions, King Puck in Killorglin, Ireland; the Kelpies-Falkirk, Scotland; the Rhinos-New York City and the mythical Eagle-Kerala, India, to name but a few, and Ljubljana is no different, it too has animals at its heart.
- We know of the Dragons and the Rat, but only a few know about the Elephant!
- There is a Hotel which marks the spot where the first Elephant was seen in Ljubljana. The Slon is so named after this first Elephant, whose name was Suleiman.
- Now Suleiman was a gift to Maximillian II on the occasion of his wedding in 1552.
- He lived a perfect life for his remaining years on Earth, but his death was not so idyllic.
- He was sent to the taxidermist and was displayed for a few hundred years, but his less-than-honourable demise finally came after being sold off for leather goods following World War II.
- We know that in Preseren Sq., the church is red because that is the colour of the Franciscan Order! But did you know?
- The Order apparently planted trees out front to shield the statues' bare bosoms from the public's prying eyes!
- Dragon Bridge was built as a 'better option' because if it had failed, had it been made in Venice (which was the option), it could have had significant consequences. Why? Well, its design and make-up

273

were untested before, so it was thought better to try it somewhere else first. Looks like a good experiment to me!

- In a twist, a house used by the Slovenian resistance in World War II is now part of the Germans Embassy

Patricks Irish Pub

Address: Prečna ulica 6, 1000 Ljubljana
Phone: +386(0)12301768
Hours:

Sunday, Monday	Holiday	Closed
Tuesday, Thursday		7:00 pm–12.00 am
Friday and Saturday		7:00 pm–1:00 am

Another little pub (No Weatherspoon's here, thank God!), with heaps of beers and atmosphere, well worth a stop off for the craic!

Route: Turn left out of the pub and follow Precna Ulica till the junction with Trubarjeva Cesta. Here we turn right and follow the road until we see the pathway Za Creslom on your left. Take this path and turn left at the bottom, Fany and Mary down here on the left.

Christmas Quote:

"That is what Christmas should be about, I think—togetherness and playfulness. It's like a game."

– Billy Howle

More Christmas Facts: Germany

Make gingerbread cookies

- Here we go again: this time, it's the Battle of the Gingerbreads!
- Nuremberg is considered by some as the Mecca of Lebkucken (gingerbread), with every bakery having

and protecting its own 'traditional, unique and authentic' recipe.

- One thing with Germany is the plethora of Christmas markets (considered by some to be the best), and what is a staple treat at them? Yes, it's gingerbread.
- The tradition of Christmas markets is widespread throughout German towns and cities. These draw marketgoers from the local regions and around the world.
- Christmas Music, Lights, Gifts, Aromas, Gluhwein and food! Love a German sausage at Christmas.
- What differs here from England is that Christmas Eve, (known in Germany as Heiligabend) is when people give out presents. Part of the tradition is to serve a small plate of potato salad with frankfurters or similar.

Finland

Family sauna!

- So on Christmas Eve, where do you go? Party/Pub? Restaurant? Not in Finland. Whole families make tracks to their nearest Sauna.
- Saunas are a prominent part of Finnish life, so much so they even have a name for the Sauna for this Christmas tradition, Joulusauna, which has been part of their culture for centuries.
- The idea behind this is to cleanse the body and soul before the festivities start. It's like having a clean plate at the start of a meal.

- They also use aromatic (essential) oils and Christmas lanterns to make it more festive. They leave a special treat for the Sauna Elf (Saunatonttu).
- It may seem unusual for other cultures to strip naked and relax in front of Mum and Dad, etc. etc. But nudity is not an issue in Finland, and indeed wearing any swimsuits or towels in a sauna can be counterproductive and lose the benefits it offers.

Poland, Germany and Ukraine
Spiders Webs

- Although spiders and cobwebs (seen by some as scary and spooky) are usually linked with All Hallows Eve (Halloween), in some parts, they are seen as good luck and form part of their Christmas tree decorations.
  - o   NB. Even in our history, spiders have played their part.
- So in Scotland at the start of the 1300s, things needed to be more settled. However, Robert the Bruce was crowned King of Scots in 1306. Robert eventually led Scotland to victory over the English in what has become known as the first war of Scottish Independence.
- One can see why this fellow is such a hero in Scotland, especially in 2022 when the Scots are pushing yet again for independence.
- However, things could have started better from the perspective of King Robert.

- You know how they say, 'out of adversity comes opportunity' Benjamin (Benjamin Franklin). Well, spookily enough, these bad times are where the spider comes into play!
- The Brits (Longshank) had defeated the Scottish army on numerous occasions. After the last battle, poor old Bruce was forced into hiding (some say in a cave or an old cottage) and while lying there contemplating his future and what would happen next. His eyes notice a little spider climbing up and trying to swing across the entrance to spin a web. However, every time he tried, he failed and fell to the ground.
- But did he give up... no! After six times, he finally made it across and spun his web. Robert saw this as an Omen, as he had been defeated by Longshanks six times. The spider gave him the impetus to try again, and in fact, on the seventh time of asking at Bannockburn in 1314, he won, which led to their independence in 1328.
- It is also believed that the phrase, "if at first, you do not succeed, try and try again," stemmed from here... but who really knows for sure. But hey! It adds the legend. I remember being told as a kid that when the English came to look for Robert, they saw the web and immediately thought that it would not be there if he had gone inside, and so left to continue their search elsewhere; well, that was lucky then.
- But back to Christmas and Europe: According to their 'Legend of the Christmas Spider', in times of yore,

there was an old widow, who was very poor, but did her best and worked hard. Unfortunately, she could not afford to decorate her yuletide tree this year. But as she slept, St Nicholas sent spiders to throw their webs over the tree. When she arose the next day on Christmas morning, the sun broke through the windows and appeared to give the webs a tinge of gold/silver. This legend is also believed to be where the inspiration for tinsel comes from.

Christmas Quote:

"There are those who give with joy, and that joy is their reward."

– Kahlil Gibran

Okay, let's go find a beer!

Fany and Mary

Address: Petkovškovo nabrežje 23, 1000 Ljubljana, Slovenia

Phone: +386 5 223 52 23
Hours:

Monday	—	11:00 am–10:00 pm
Tuesday	—	11:00 am–10:00 pm
Wednesday	—	11:00 am–10:00 pm
Thursday	—	11:00 am–10:00 pm
Friday	—	11:00 am–10:00 pm
Saturday	—	11:00 am–10:00 pm
Sunday	—	11:00 am–10:00 pm

Great location by the river, close to the bridges (another bridge with padlocks; they seem to pop up everywhere), perfect for watching the world go by. When I visited, we had lovely food, and the staff here was very friendly and attentive. If you are visiting with your dog, this place is very accommodating. Stunning location and has charming and polite servers. We had great pizza and classic burgers.

Route: Out of the pub, cross over the Mesarski most (Padlook Foot Bridge), carry on straight into Dolnicarjeva Ulica, then turn left into Ciril Metodov Trg and follow until you see Ribja Trg on your right (Just after Stritarjeva Ulica). As you emerge into a little square, the Ta Bar is on your right.

As you cross the bridge, there is something unusual.

Can a river have more than one name?

Well, in Slovenia, it certainly can. As the river meanders its way across the country, it disappears and rises again further down. Not realising it was the same river, locals named it as it appeared where they lived. Only later on was it found to be one and the same.

Its known names are Trbuhovca, the Obrh, the Stržen, the Rak, the Pivka and the Unica rivers, then the Ljubljanica as it rises in the city.

The bridge itself is known as Butcher Bridge (Lovers Bridge), as lovers come here and padlock their love to the bridge as a sign of commitment and continuity. Built in 2010 at the cost of nearly 3 million euros. However, the bridge was initially considered in the 1930s but never happened because of WWII.

On your way, you may see a strange sight, a kangaroo statue!

No this is a water fountain.

Address: 1000 Ljubljana

Ljubljana is an anomaly in that it is proud of its freshwater supplies. It promotes it fully, recommending people fill up their bottles etc., from the fountains placed around the city.

What is fun about this, they are not the typical run-of-mill taps; no, they are an assortment of different shapes and sizes (even the kangaroo above is one). If you have time, download the Tap Water Ljubljana app, which provides a map of all of them and locates your nearest one based on your GPS signal.

As a separate trek, trying to find them all and collect the photos may be worthwhile, but for us now, let's get on with some beer.

As you go along, you will see St Nicolas Cathedral on your right; check it out and make sure you look at the doors.

Pope John Paul II's visit to the Cathedral made quite an impression.

Before his visit in 1996, new doors were added to the front and the side. These bronze doors are sculpted and beautiful to look at. One has the figures of six Bishops, and the other is the Slovenian history of Christianity.

As you walk down the road, look up to your left, and you may glimpse the Castle; here are some things you may not know.

City of Dragons!

- Dragons are associated with Ljubljana via history/legends and maintain prominence in the city in the form of The Dragon Bridge. (A bit further down past Butcher's Bridge.)
- If you are into flags etc., the image of the Dragon is a mainstay of Ljubljana's coat of arms.

- So where did it all start?

  o So, you must have heard or remembered from mythology the tales and exploits of Jason, his ship, the Argo, and his crew, the Argonauts.

  o After he was ordered by King Pelias, who usurped his dad, Jason was sent on a mission to retrieve the Golden Fleece from another King.

  o So he managed to win the fleece against odds (Oh, and by the way, the King's daughter who fell in love and fled with him!), and while making his escape, his ship sailed up the Danube, along the Sava, finally reaching the Ljubljanica and Ljubljana.

  o Here they decide to take the Argo apart, carry it across the land to the Adriatic Sea, and then rebuild it and sail home to Greece.

  o But winter was about to set in, so they set up camp where the River Ljubljanica rises until winter broke.

  o But on this marsh, here there be dragons!

  o You could guess the rest of the story, even if you had never heard of it. Hero… camping on land that the Dragon believed was his…

  o So a fight ensued, and the Dragon retreated into his cave (which was under where the Castle now lies), leaving Jason as the victor.

- It is also said the Dragon still lives there today, and when he is unhappy, the ground shakes.
- By slaying the beast, Jason opens up the land to be free from dragons and settle. And that's Ljubljana.
- There is more to the slaying of the Dragon, involving spells and potions from a sorceress (Jason's girlfriend!), but I will let you find out if you are interested at your leisure.
- Finally, it is also believed that the Dragon on the bridge does stir, occasionally twitching his tail, but this is almost invisible! What could cause this to happen? Well, on the rare occasion when a chaste and pure virgin walks across the bridge, the Dragon seems to like this and moves his tail in appreciation.
- Now whether it is muscle memory from all sacrifices of virgins the legends allude to, to appease the Dragon, or whether he likes women, who knows.

- Now there is another version, which may have more credence.
- This again involves the Castle but in a different, more religious way.
- Back in 1489, the Castle's Chapel was linked with Saint George, yes, you know, the one, the dragon slayer.
- So here is where the tenuous link appears; stay with me on this one.

- As the Castle looks over the town, and the Chapel has the patronage of Saint George, he is looking after the Chapel, and the Castle.
- So Saint George takes care of the Chapel, and the Chapel takes care of the Castle, and the Castle takes care of the City. Are you still with me?
- So as St George is synonymous with the Dragon, it is only a small leap of faith to see how the two changed places & the Dragon became the symbol of Ljubljana.
- Now there is another legend associated with the Castle, and that is of Frederick the Rat (we will come to rat a bit later).
- So Frederick II (1379–1454), apparently fed up with his original wife, murdered her (allegedly), so he could marry another, whom his father did not approve of.
- Not in a forgiving frame of mind, the father (allegedly) had the new wife drowned while imprisoning Frederick in the Castle's tower.
- He only survived the ordeal when a servant excavated a secret tunnel which allowed him to smuggle in some food and kept him alive.

Now Frederick the Rat is the castle rat! His heroic adventures saved Ljubljana and the Castle from the Turkish invasion. You can purchase a copy from the Castle's souvenir shop.

But a little more about him.

Frederick is a smart rat who is reputedly more than 500 years old. Taking his name from Emperor Frederick III of Habsburg.

Emperor Frederick was the first ruler to build a fortress on what we know now as Castle Hill.

But Frederick lived there before the Castle was built and, in fact, was probably the first known resident of the Castle.

Although he had the run of the Castle, his preferred rooms were in the Erasmus Tower.

As normal rats were commonplace, no one took too much notice of him; a bit strange that he walked on two legs, but hey! He was just a rat!

He would wonder about the whole Castle listening in on conversations and watching events as they unfolded, also being just a 'rat', prisoners, who had no one else to talk to, would confide in him. Imagine if you could have a chat with him (as he is still there today) and the stories he could tell.

Christmas Quote:

"A toy is never truly happy until it is loved by a child."

– Rudolph the Red-Nosed Reindeer

Onwards to the pub.

Ta Bar

Address: Ribji trg 6, 1000 Ljubljana, Slovenia
Phone: +386 31 764 063

Hours

Monday	—	5:00 pm–12:00 am
Tuesday	—	5:00 pm–12:00 am
Wednesday	—	5:00 pm–12:00 am
Thursday	—	5:00 pm–12:00 am
Friday	—	5:00 pm–12:00 am
Saturday	—	5:00 pm–12:00 am
Sunday	—	Closed

This bar, just a stone's throw from the River, is a lovely little hideaway which packs a surprising punch. Now, not the cheapest, but the ambience and food are lovely and a nice selection of wines to boot. The staff are very friendly and helpful too.

As this is the last bar on our crawl, here is a titbit which I feel adds a nice ending to the trip.

If you are still unaware (maybe you have read this book out of order), I was for most of my life a London Firefighter. You can imagine my glee when I realised in Ljubljana there was a St Florian's Church, now St Florian is the patron saint of firefighters, so it was a must that I visited it. The map at the start of this crawl shows where it is sited, and here are some bits you may find interesting.

St Florian's Church

St John of Nepomuk

Now here are some interesting facts about St John.

- Saint John is also prominent in Prague, Ljubljana and other parts of Europe.
- He retains a special place in my heart as he was one of the first legends I unearthed many years ago, an ex-firefighter; this also explains why it is in this book, both in Bruges and here in Ljubljana.
- You can read his story in the Bruges crawl, but why do you ask should I carry on with the tale here? That is because it's also here. When I visited St Florian's Church (patron saint of firefighters), the depiction of

St John's death/murder is a centrepiece on the outside of the Church.

So we know about St John, but who was St Florian.

Where to start, well, the statue itself depicts St John, but John has something in common with St Florian… read on.

So how did Saint Florian become the patron saint of firefighters and seen as a protector from water and fire?

- As an Officer of Rome in their army, Florian Von Lurch eventually died by refusing to deny his faith. Indeed if the stories are to be believed, his life and death were anything but conformist.
- When Florian lived in the third century (AD 250–304), Christians were not really respected by the Romans. In fact, it was during the Great Persecution of the Christians by Rome that eventually saw the demise of Florian.
- Florian rose through the army ranks and ended up as a Commander.
- Commander Florian was also supposed to have led and trained a special group of soldiers in the art of firefighting.
- Florian was charged with 'disciplining' the Christians, and when he refused, despite being cajoled by another of Emperor Diocletian's men (Aquliius), Florin still refused to persecute them and so was flogged.
- But Florian was still unrepentant, and, Aquliius panicked and felt that Florian's 'bravery or bravado'

would stimulate others to rebel against Roman Empire.

- It was at this point Aquliius gave the order for Florian to be burnt alive.
- But Florian was not yet done!
- In the face of adversity, he taunted Aquliius, so much so that he decided not to burn him but have him tortured and thrown into the Enns River with a rock secured around his throat, whereupon he drowned.

Rome gave St Florian the title of the patron saint of firefighters and floods.

So we now know why he is the patron saint of firefighters but is that enough?

Along with the above, another legend states that when faced with a large conflagration that threatened a whole village, Florian extinguished all the flames with one small bucket of water.

He is also linked with water and drowning, as he drowned himself.

So in the immortal words of Star Wars, May the Force be with you?

Why, well, St Florian's feast day is 4 May!

End of the Road!

So if you managed to work through the six Christmas markets and all the pubs therein, you should congratulate yourself.

To say this has been a 'chore' would be wrong. I have loved compiling/writing this book and researching all the pubs, history, folklore, and Christmas!

While I Have always prided myself on being 'tuned in' to Christmas, I have learned much more during the book's writing.

So what can I say about Christmas?

As a Catholic, I believe in the 'true meaning of Christmas', which should never be forgotten. Yet I feel there is also a need for a bit of 'imagination' and belief in something 'magic' for children to look forward to, and a belief in someone who cares about children.

This is summed up by the lawyer in Miracle on 34th Street when he tells the judge: "I ask the court to judge which is worse: A lie that draws a smile or a truth that draws a tear."

Miracle on 34th Street (1994), IMDb

Children grow up so quickly nowadays. Surely having them use their imagination, or just looking forward through children's eyes, for the whole season of Christmas cannot be wrong.

Well, not in my books.

Christmas for me is a time of giving, a time when we get together with our loved ones, celebrate our lives, remember those who are no longer here, and tell the tale of Mary, Joseph and Jesus, Little Donkey, the Magi, as well as Father Christmas and his elves.

Well Done!

References

"Pepys's Diary," Guild Publishing, London, 1981, ed. Robert Latham, 107–8.

80 Best Christmas Quotes, Merry Christmas Quotes and Sayings. *https://www.today.com/life/holidays/christmas-quotes-rcna43138*

105 fun facts about Christmas to impress your friends, Tomango. *https://www.tomango.co.uk/blog/105-fun-facts-about-christmas-to-impress-your-friends/*

Sources and Further Reading

When writing this book, I have tried to verify the knowledge I have picked up over the years, and to do that, I researched the internet (a lot easier than going to a library, that's progress for you), as with the web, links take you to different links and before you know it you are deep in reading.

To that end, should you wish to dig deeper, here are the websites I have used.

Having spent much of my life gathering trivia, some of this book is accurate, but as with folklore, legends. Myths etc., and stories vary over time.

I have also visited all the cities and engaged with the locals in the best way possible to glean the best pubs, bars, and local stories.

Please read on if I have planted a seed in your mind that needs nurturing.

https://www.stratfordplay.co.uk/about-us
https://www.spookyisles.com/stratford-upon-avon/
https://stratfordobserver.co.uk/news/researching-stratfords-historic-buildings-38299/
https://britishfoodandtravel.com/2020/06/26/britains-historic-pubs-part-1/

https://www.cotswolds.info/places/stratford-upon-avon/interesting-facts.shtml

https://www.etymonline.com/word/jacobean

https://www.historyhit.com/the-bizarre-life-of-cats-in-shakespeares-england/

https://museumhack.com/shakespeare-facts/

https://www.shakespearesglobe.com/discover/shakespeares-world/the-globe/

https://nosweatshakespeare.com/resources/shakespeare-facts/

https://www.visitestonia.com/en/why-estonia/the-tale-of-tallinns-most-famous-christmas-tree

https://arhivs.rigasnami.lv/en/news/view/magical-and-historical-story-about-christmas-tree-and-the-brotherhood-of-black-heads

https://www.bellshakespeare.com.au/common-phrases-in-shakespeares-plays

https://inews.co.uk/culture/william-shakespeare-quotes-phrases-plays-465th-birthday-today-3210

https://sites.google.com/site/majorriversofthebritishisles/river-avon?pli=1

https://kids.kiddle.co/River_Avon,_Warwickshire

https://www.visitstratforduponavon.co.uk/attractions/bancroft-gardens

https://statues.vanderkrogt.net/object.php?record=gbwm081

https://www.shakespeare.org.uk/

https://englishhistory.net/middle-ages/richard-i/

https://charlenenewcomb.com/2014/12/20/pirates-shipwreck-and-the-capture-of-a-king-december-1192/

https://www.historyhit.com/facts-about-richard-the-lionheart/

https://visitbath.co.uk/things-to-do/pulteney-bridge-p56151

https://visitbath.co.uk/things-to-do/beazer-maze-p56481

https://visitbath.co.uk/blog/read/2021/01/fifteen-little-known-facts-about-bath-b68

https://www.mysteriousbritain.co.uk/hauntings/the-garricks-head-bath/

http://www.ghost-story.co.uk/index.php/haunted-pubs/148-the-garrick-s-head-inn-bath-england

https://www.dailymail.co.uk/news/article-399017/The-steamy-truth-Roman-Bath.html

https://www.discoverwalks.com/blog/rome/top-10-fun-facts-about-the-roman-baths/

https://www.bathnes.gov.uk/services/environment/bath-hot-springs

https://edsimoneit.blog/2019/03/30/the-christkindl-gift-bringer-tradition-was-first-introduced-by-martin-luther/

https://www.stcatherinercc.org/single-post/2020/01/01/where-do-we-get-the-names-of-the-three-magi

https://en.wikipedia.org/wiki/Christmas_market

https://stpaulchristmasmarket.org/history/

https://www.looper.com/287617/things-only-adults-notice-in-its-a-wonderful-life/?utm_campaign=clip

https://www.hrp.org.uk/tower-of-london

https://www.theravenofbath.co.uk

https://www.bathguildhallmarket.co.uk/

https://www.museabrugge.be/en/visit-our-museums/our-museums-and-monuments/belfort

https://www.holyblood.com/homepage-of-the-basilica-of-the-holy-blood

http://www.worldholidaytraditions.com/Countries/Belgium

https://scarymatter.com/2021/09/09/bruges-haunted-house/

https://www.bbc.co.uk/news/uk-england-kent-16834429

https://great-castles.com/rochesterghost.html

https://medwaymemories.co.uk/picturesque-love-lane-haunted-by-half-a-ghost/

https://www.thehistorypress.co.uk/articles/a-history-of-british-pub-names/

https://www.joseph-holt.com/news/history-of-pub-names

http://scribblinglau.blogspot.com/2018/02/unusual-things-see-do-in-rochester-kent.html

https://www.theguardian.com/world/2009/mar/12/race-monarchy

https://geofframbler.blog/2018/10/23/spymaster-of-rochester-from-1908-1913/

https://www.royal.uk/queen-charlotte

https://thetudorenthusiast.weebly.com/blog/the-execution-of-bishop-john-fisher

https://www.britishmuseum.org/collection/term/BIOG11 5655

https://www.rochestercathedral.org/gardensblog/history

https://city-of-rochester.org.uk/features/hidden-rochester/

https://www.thepilgrimsway.co.uk/

https://city-of-rochester.org.uk/features/hidden-rochester/

http://www.dover-kent.com/2014-project-c/George-Vaults-Rochester.html

https://read.amazon.co.uk/?asin=B00IC93UGC

https://www.stratfordplay.co.uk/about-us

https://www.kasteelvalkenburg.nl/en/

https://www.christmastownvalkenburg.com/to-do/christmas-market-municipal-cave/

https://www.atlasobscura.com/places/kasteelruine-fluweelengrot

https://www.castles.nl/valkenburg-castle

https://en.wikipedia.org/wiki/Valkenburg_resistance

http://www.aachen-webdesign.de/verzet/bevrijding.php?lang=en

https://www.pudforallseasons.com.au/blog/christmas-pudding-history-and-traditions/

https://www.countryfile.com/go-outdoors/days-out/top-10-quirky-christmas-traditions/

https://www.christmastreeworld.co.uk/blog/48-christmas-facts-statistics-you-probably-didnt-know/

http://www.christmasfacts.net/

https://oceanwide-expeditions.com/blog/st-nick-and-the-arctic-the-north-pole-christmas-connection

https://www.lightailing.com/blogs/news/lego-latest-usage-and-revenue-statistics

https://celadonbooks.com/6-things-you-didnt-know-about-the-night-before-christmas/

https://www.ducksters.com/geography/country/netherlands_history_timeline.php

https://www.history.com/news/why-is-christmas-celebrated-on-december-25

https://historicengland.org.uk/listing/what-is-designation/heritage-highlights/did-oliver-cromwell-really-ban-christmas/

https://www.bbc.co.uk/news/uk-wales-35120354

https://www.ljubljanskigrad.si/en/castle-events/Friderik-the-Castle-Rat/

https://www.timeref.com/places/bath_abbey.htm

https://www.nationalgeographic.com/culture/article/are-french-fries-truly-french

https://www.medway.gov.uk/info/200242/christmas_in_medway/643/rochester_christmas_market

https://www.rochesterdickensfestival.org.uk/index.htm

https://www.kerstmarktgemeentegrot.nl/en/home

https://www.christmastownvalkenburg.com/to-do/christmas-market-municipal-cave/

https://www.bbc.co.uk/news/uk-england-kent-16834429

https://inews.co.uk/culture/television/blackadder-quotes-161437

https://www.rochestercathedral.org/archive/orchard

https://www.historyextra.com/period/stuart/a-king-without-a-crown-james-iis-years-in-exile/

https://www.pepysdiary.com/diary/1665/06/

https://www.thoughtco.com/shakespeares-death-facts-2985105

https://nosweatshakespeare.com/resources/shakespeare-facts/

https://www.imdb.com/title/tt0110527/characters/nm0001518

https://www.poetryfoundation.org/poems/44621/account-of-a-visit-from-st-nicholas

http://bath-heritage.co.uk/nelson.html

https://www.independent.co.uk/arts-entertainment/books/features/st-patricks-day-2018-flann-o-brien-third-policeman-irish-humour-ireland-books-writing-a8252936.html

https://poets.org/poem/raven

https://www.litcharts.com/shakescleare/shakespeare-translations/a-midsummer-nights-dream/act-2-scene-1

https://www.richardwatts.org.uk/st-catherine-s-hospital

http://www.dover-kent.com/2014-project/Eagle-Tavern-Rochester.html

https://nosweatshakespeare.com/blog/shakespeare-and-christmas/

https://staging.myshakespeare.com/taming-of-the-shrew/introduction-2

https://dirkdeklein.net/2016/03/08/forgotten-history-pierre-schunck-resistance-fighter/